On-Board Design Models and Algorithm for Communication Based Train Control and Tracking System

Railway systems have a long history of train protection and control, as to reduce the risk of train accidents. Many train control systems include automated communication between train and trackside equipment. But several different national systems are still facing cross-border rail traffic. Today, trains for cross-border traffic need to be equipped with train control systems that are installed on the tracks.

This book covers the latest advances in Communication Based Train Control System (CBTC) research in on-board components locomotive messaging systems, GPS sensors, communications wayside and switching networks. It also focuses on architecture and methodology using data fusion techniques. New wireless sensor integrated modeling techniques for tracking trains in satellite visible and low satellite visible environments are discussed. With a Tunnel Surveillance Integration model, the use of optimal control is necessary to improve train control performance, considering both train–ground communication and train control.

The book begins with the background and evolution of train signaling and train control systems. It introduces the main features and architecture of CBTC systems and describes current challenges, methods and successful implementations.

This introductory book is very useful for signal and telecommunication engineers, enabling them to get better acquainted with the technology used in CBTC and help them implement the system suitable for Indian Railways. As this is a new technology, the information provided in this book is generic, and will be subsequently revised after gaining further experience.

Power Electronics and Applications Series

Series Editor:
Muhammad H. Rashid
University of West Florida, Pensacola, Florida USA

Complex Behavior of Switching Power Converters
Chi Kong Tse

DSP-Based Electromechanical Motion Control
Hamid A.Toliyat and Steven Campbell

Advanced DC/DC Converters
Fang Lin Luo and Hong Ye

Renewable Energy Systems: Design and Analysis with
Induction Generators
M. Godoy Simoes and Felix A. Farret

Uninterruptible Power Supplies and Active Filters
Ali Emadi, Abdolhosein Nasiri, and Stoyan B. Bekiarov

Modern Electric, Hybrid Electric, and Fuel Cell Vehicles:
Fundamentals, Theory, and Design
Mehrdad Eshani, Yimin Gao, Sebastien E. Gay, and Ali Emadi

Electric Energy: An Introduction
Mohamed El-Sharkawi

Electrical Machine Analysis Using Finite Elements
Nicola Bianchi

For more information about this series, please visit: https://www.routledge.com/
Power-Electronics-and-Applications-Series/book-series/CRCPOWELEAPP

On-Board Design Models and Algorithm for Communication Based Train Control and Tracking System

Tanuja Patgar and Kavitha Devi CS

CRC Press
Taylor & Francis Group
Boca Raton London New York

CRC Press is an imprint of the
Taylor & Francis Group, an **informa** business

First edition published 2022
by CRC Press
6000 Broken Sound Parkway NW, Suite 300, Boca Raton, FL 33487-2742

and by CRC Press
4 Park Square, Milton Park, Abingdon, Oxon, OX14 4RN

ISBN: 978-1-032-27724-0 (hbk)
ISBN: 978-1-032-27774-5 (pbk)
ISBN: 978-1-003-29401-6 (ebk)

DOI: 10.1201/9781003294016

Typeset in Times
by codeMantra

I owe my deepest gratitude to my family. Their constant inspiration and guidance has kept me focused and motivated. I am grateful to my father, **Mr. Parameshwar,** for teaching me to be independent and hardworking. My gratitude to my mother, Late **Saraswathi,** whose unconditional love has been my greatest strength, cannot be expressed in words. My sincere thanks go to my husband, **Mr. Harish,** for his understanding and love during the research work. His support and encouragement were in the end what made this dissertation possible. My daughter, **Prachi,** has been always my pillar my joy and my guiding light.

Dr.Tanuja patgar

I whole heartedly remain grateful to my son **Hemachandra Suri,** husband **Suresh Kumar G**, parents **Lakshmi Devi M & Srinivasa J**, my sister **Asha Devi CS** and brother **Sanjay Kiran CS** for their continuous support throughout my research work.

Dr. Kavitha Devi CS

Contents

Preface

MOTIVATION FOR THE BOOK

Innovations in the field of Communication and Information technology is boosting a new era of wireless-based solution. Today many global sectors have identified potential areas to be improved with the effective exploration of Information and Communication Technology (ICT). Rail transportation is one such sector that can be improved to provide safe and efficient service to the general public with the integration of advanced wireless communication and sensor-based technology.

Many research studies have been done so far in wireless communication and networking technology applied to the railway. This challenge drastically changes the path that we dreamt of transportation system of rail and its supporting system in the way we drive in the future. To identify the rail transportation safety issues in some notable cases such as accidents due to collision, at level crossing, derailment due to over speed, track and signaling fault, frequent train delay due to landslides are minimized only by continuously monitoring the train. Accurate estimation of the train's position and speed are the safety measures needed for automatic train operation. None of the solutions is possible unless the existing train tracking and controlling system are replaced by a more informative feasible system that showcases accurate train position and speed of the entire network.

The train tracking and control system are said to be automatic as well as intelligent systems. But these have limitations such as they comprise too many components, depend on traditional track circuit, infrastructure failure, traverse incorrectly lined switch, wayside switch fails to detect position and fail to convey message, on-board system failure, insufficient braking force to de-accelerate the train and missing speed limit in database. Hence, the use of innovative technology with a powerful solution is essential for revolution in the modern railway sector.

One such technology that continues a heritage of innovation with the next generation in finding more accurate location data is Global Positioning System (GPS). It offers many benefits over traditional track circuit to both manufacturer and end user. Hence, leading the revolution in the remote monitoring market, based on GPS technology, wireless communication, signal processing and embedded computing have become inexpensive and relatively compact. GPS-based system has significant in reducing accident, delay, to improve signal capacity and cost-effectiveness of transportation.

The GPS-based on-board train tracking and controlling system with secure wireless data communication network for continuously monitoring the train location, speed and running direction accurately. It is important that the future system should deliver robust performance under an adverse environment condition of high radio jam noise, offer low life-cycle cost and is able to adjust on any existing system to facilitate replacement. The tracking model has also pitched the way for locating predefined place based on the current moving object. The object location mapped with other information-loaded server to locate nearby service. Besides the diverse application of GPS-based tracking models, the reliability of service is still limited. GPS-derived positioning information accuracy is dependent on surrounding conditions as well.

PHILOSOPHY

Different countries have developed their own train control system for their railroad. Positive Train Control System (PTCS) and Communication Based Train Control System (CBTC) are developed by American. European people have implemented European Train Control System (ETCS), while the Chinese Railway created Chinese Train Control System (CTCS) and East Japan Railway produced East Japan Train Control System (EJTC). The system is different from one another in tracking, data transmission technology, integration of new component as well as simulation methodology.

Indian Railway has remained adaptive to new technologies in the field of communication and signaling since 1853. The introduction of relay interlocking, various block working, multiple aspects of signaling, train detection and radio communication technology etc., was adopted by other railways across the globe. However, last 10 years, the deployment trend of imported or domestic new technology-based system and equipment of the Indian Railway have not been very encouraging. The LED-based color Light Signal Unit (LSU), Integrated Power Supply System (IPS), Single and Multiple Sections Digital Axle Counters (DACO), Solid State Inter-locking, ETCS (European Train Control System) level II-based Train Protection and Warning System (TPWS), GPS based Anti-Collision Device (ACD) and GSM (R) based train radio communication system have been introduced.

The book is a design tracking system that shows improvement in performance of train detection and tracking accuracy in two different environments

(1) Satellite Visible Environment (2) Satellite Low Visible Environment. As the system localization accuracy degrades when it passes through GPS denied areas such as bridge, dense forest, tunnel, mountain, valley, slope, etc. due to a reduction in signal quality and it suffers from Line of Sight (LOS) and Multi-Path Loss (MPL).

Hence an accurate and efficient continuous tracking performance at the core of such a system is essential for building higher level Internet of Things (IOT)-based intelligent solution. The following is our vision toward building a train tracking system in real-time application. (1) Tracking train in satellite visible and low satellite visible areas with precision and accuracy. (2) Positioning Rail Accurate Communication Highway Identification system is designed with Wireless Sensors (MEMS-based sensor Accelerometer and Sonar Ranger), Radio Frequency Identification (RFID) and Differential Global Positioning System (DGPS). (3) The sensor-based On-board train tracking and control system is designed to get tracking accuracy of 95%–98%. (4) The kinematics Update State Hypotheses Information Surveillance system is designed for tracking and monitoring the moving train in satellite visible environment (Outside the tunnel). (5) Heterogeneous Access Remote Integrating Surveillance Heuristic tracking system is designed for tracking the moving train in tunnel. (6) Installing expensive track circuit, wayside equipment, level crossing communication interface unit and building additional communication infrastructure will no longer be needed.

The book proposed the algorithm based on Discrete Kalman Filter (DKF) for robust train tracking in satellite visible area. We have selected Discrete Kalman Filter methodology for kinematic (position, velocity, speed, direction and acceleration) measurement of the moving train. It takes care of missing and noisy measurement and continuously provides the best estimation depending on the available measurement value. The proposed method has been shown to effectively model the errors associated with Differential GPS measurement. However, even after the tracking measurement error compensation, there exists residual affecting the standalone DGPS tracking accuracy.

The integration methodology can adopt for Differential GPS error estimation through WSN and RFID. Further, it is compensated to derive a highly accurate and reliable tracking solution when the train is moving in tunnel. The existing methodology demonstrates the limited performance of low-cost DGPS-WSN-RFID integration systems due to their inability to capture highly non-linear tracking accuracy error variation. Our book focus on implementing intelligent techniques such as wireless tracking controller based on Quadratic Optimal Control theory is proposed. Overall performance of the control design is based on Liapunov approach, where Quadratic Performance Index is directly related to Liapunov functions. By minimize and maximize the Performance Index value corresponding to control input will trace the tracking error inaccuracy. The outcome of our book is expected to overcome the limitation associated with existing intelligent techniques to model the low-cost sensor-based tracking accuracy.

Recently, statistical-based approaches are gaining popularity in engineering community for handling the highly linear input-output functional relationship effectively. Thus, an initiative is taken to demonstrate the applicability of statistical-based techniques to fuse the low-cost Wireless Sensor, RFID and Differential GPS. We also focus our attention on considering only two sensor combinations and test the result. To do this, we proposed a data fusion algorithm using Square Root Information filter. It is attempting to gather information from Differential GPS and Wireless Sensor. The proposed approach intensifies the Di-filter model-based tracking algorithm called Interacting Multiple Model (IMM) algorithm is designed to track the train kinematic update in Satellite Visible and Low Satellite Visible areas. The suggested algorithm is reduced the root mean square error in velocity and position measurement. We have implemented the Distributed architecture for data fusion and classified sensor measurement fusion methodology such as Differential GPS-WSN-RFID fusion system based on Bayes probabilistic model, Differential GPS-WSN fusion system based on Information Filter Least Squares Technique and DGPS-DGPS fusion based on Di-Kalman Filter with IMM algorithm.

In summary, we have designed the train tracking system using DGPS-WSN-RFID for satellite visible and low satellite visible environment. We have successfully implemented the proposed scheme using data fusion algorithm. The results of experiments suggest that tracking inaccuracy estimation error of 0.02%, when the train is moving in a satellite visible environment (model1) and comparable tracking inaccuracy estimation error shows 0.37% when the train is moving in a poor satellite visible environment (model 2). This study proves the data fusion model offers a significant detection performance accuracy level of 95%–98% for moving train across a wide range of operational scenarios.

SUMMARY BY CHAPTERS

Chapter 1 presents a brief introduction on major segments that are used in each tracking and controlling system. We also present additional information on how these technologies are helpful in solving real-world railway system. With the development in data communication, computer tool and control technique, automated rail control system such as Communication Based Train Control System (CBTCS), European Train Control System (ETCS), Chinese

Train Control System (CTCS) and Indian Train Tracking and Control System Scenario are presented in this chapter.

Chapter 2 covers the integration systems utilized in the implementation of an intelligent train monitoring system using different sensors such as RFID, GPS, GPRS and GIS. Developing the theoretic framework on multi-sensor data fusion is taking context into consideration. The method is to combine numerical and symbolic information, in order to have a fusion process in different levels. The level of treatment that analyzes the context data using contextual variables for the estimation process model is designed. The outcome result is dominating the measurements provided by sensors well-adapted to maximum context data and to minimize the not well-adapted sensors.

The principle behind each positioning system using Differential GPS, RFID and Wireless Sensors is explained in **Chapter 3**. It also emphasizes the importance of DGPS-WS-RFID integrated system to improve accuracy, continuity and reliability of the tracking solution. It enhances on how standalone system solution is compensated with integrating system and highlight some of their advantages and limitations.

Additional information on tracking surveillance algorithm and relevant software simulation used for implementation is presented in **Chapter 4**. We also briefly discussed the advantages of using Discrete Kalman Filter in solving real-world problems and WSN-RFID-based train monitoring system. Finally, the chapter also presents a sensor model matching control system, tracking controller design, followed by Quadratic Optimal Controller design based on Liapunov approach.

Chapter 5 discusses the design of proposed *"Sensor Accuracy Remote Access Surveillance Wireless Automatic Tracking Heuristic Innovation"* *(SARASWATHI)* model based on DGPS satellite-based navigation system. The improved trajectory tracking accuracy of train in SVE and SLVE is maintained. Di-filter model analysis is proposed with IMM-based tracking algorithm IMM that makes tracking system to better match changing target dynamics as well as the current situation of the train. The balancing method of the Di-filter model for two environment cases is difficult to manage algorithm computation for the required tracking accuracy.

A novel data fusion process algorithm using Square Root Information Filter is presented in **Chapter 6**. The data fusion process at the decision level is done by merging information from DGPS and Wireless Sensors for monitoring the train journey in Satellite Visible Environment and Satellite non-visible Environment. The chapter also concentrated on multi-sensor data fusion based on a probabilistic model. In this method, a single network is responsible for gathering information from many sources. The probability model is analyzed using the Bayes theorem.

Chapter 7 covers DGPS-WSN-RFID based data fusion system where fusion takes place at a higher level. Accurate estimation of train position and speed profile depending on the type of environment is required for automation. The smart rail safety system will definitely replace the existing one. The proposed method uses three sources such as DGPS, WS and RFID model for two kinds of environments such as Satellite Visible Environment and Satellite nonvisible Environment. The model analysis is described by likelihood matrix where observation and prior information are estimated in discrete form.

The primary objective of **Chapter 8** is to provide a basic understanding of machine learning methods for rail trip data analysis. This chapter discusses how machine learning methods can be utilized to improve the performance of rail trip data analytics tools. It focuses on selected machine learning methods and the importance of quality and quantity of available data. It presents a comprehensive view on these *machine learning algorithms* that can be applied to enhance the intelligence and the capabilities of an application in predictive analysis of intelligent rail trip detection service.

MATLAB® is a registered trademark of The Math Works, Inc. For product information, please contact:

The Math Works, Inc.
3 Apple Hill Drive
Natick, MA 01760-2098
Tel: 508-647-7000
Fax: 508-647-7001
E-mail: info@mathworks.com
Web: http://www.mathworks.com

Acknowledgments

We would like to express our deep and sincere gratitude to Dr. **Muhammad H Rashid** for his inspiration and for guiding us to publish this book. We express our sincere thanks to the Management of Dr. Ambedkar Institute of Technology, Bengaluru, for providing the required facilities to carry out this work.

Dr.Tanuja Patgar and Dr. Kavitha Devi CS

Authors

Tanuja P. Patgar received a BE degree in Electronics and Communication from Kuvempu University, Karnataka, India in 1996. In 2010, she received ME in Control and Instrumentation from University Vishweshraya College of Engineering, Bangalore, India. She received her PhD from Visvesvaraya Technological University, Belgaum, India in 2020. Her research field is Wireless Sensor Network, Artificial Neural Network, Machine learning, Deep Learning, Data Science and Computer Vision. She has pioneered research in wireless sensor network, control system and has published 35 international refereed Journal papers. She has edited several international and national journals, books and frequently gives invited keynote lectures at national. She is awarded national Innovative Educationalist in Engineering and Technology, Maulana Abdul Kalam Azad excellence Award of Education, Talent Teacher Ideal Teacher Pratibha Puraskar, Best paper award. She is an academic adviser for AI-based startup. She has served as a congress chair for International Congress on Mobile and Wireless Hong kong and technical program committee member and scientific committee member for IFERP and IAENG Conference. She is a fellow member of I.S.T.E, I.E.T.E, I.S.A, IMAP, IFERP, IAENG and SDIWC. Presently serving at Dr. Ambedkar Institute of Technology, Bangalore, India.

Kavitha Devi CS received BE degree in Electronics and Communication Engineering from Visvesvaraya Technological University, Belagavi, Karnataka, India in 2003. In 2010 she received M.Tech in Computer Science Engineering Visvesvaraya Technological University, Belagavi, Karnataka, India. She is pursuing her PhD in "Design and Implementation of Compact Microstrip Filters" in Visvesvaraya Technological University, Belagavi, India. Her research interest includes Wireless Communication, Microwave and Engineering, Antennas and wave propagation, Electromagnetic Waves, Wireless Sensor Network, Virtual Reality, Deep Learning, Data Science and Computer Vision with 29 international and national referred paper publications, 2 patents and 2 silver medals in NPTEL courses. She has reviewed many papers at national and international conferences and journals. She is a fellow member of I.S.T.E, C.R.S.I, and I.S.S.S & I.E.T.E, organised several technical events and hosted many programs. Presently, she is serving as assistant professor at Dr. Ambedkar Institute of Technology, Bangalore, India.

Vision of Intelligent Control and Tracking Rail System: Global Evident Data

1

1.1 INTRODUCTION

The rapid growth in the field of wireless and mobile technologies plays a key role in the global Information and Communication Technology (ICT) sector. The tracking solution of moving object is one such technology that stretches across various industries such as land, maritime and aviation. The application areas in land vehicle tracking are numerous, including fleet management, on-board navigation, stolen vehicle recovery and its enhanced services, etc. Today, it is becoming a big reality due to the strong integration of positioning technologies along with sensor integration, wireless communication and information management leading to Internet of Things (IOT)-based services.

DOI: 10.1201/9781003294016-1

This really has further pushed the market segment to avail the end user with future-generation low-cost, wireless, compact-sized in-vehicle tracking and guidance system.

The fundamental aspect of tracking technology is to estimate the time-varying position, velocity and acceleration of moving object. In Satellite Visible Environment (SVE), object tracking is easier than in the Satellite Low Visible Environment (SLVE), as it is more unpredictable and inconsistent. Hence, it is very hard for a system designer to balance the parameters in achieving satisfactory performance in terms of accuracy, power consumption, range, system implementation, cost and maintenance. In the past decade, many new technologies have emerged to achieve accurate and reliable tracking of objects in both outdoor and indoor environment. The improvement has been significant. The real-time application scale varies from many personnel objects to public transportation applications such as bus, train automation system. This tedious task is achieved by combining information from heterogeneous sensors capable of sensing vehicle's relative and absolute motion.

On the other hand, when this vision is adopted in tracking, controlling and monitoring train, it enhances the challenging innovation in the rail sector. Today, the ability to achieve business global sector's demand depends purely on availability, accuracy and reliability of information. The railway is one of the world's biggest transportation systems. The provision of a safe and reliable mode is a primary requirement of railway as more population depend on this service and prefer it as their first choice of transportation. Several constraints are taken into consideration while achieving the reimplementation of the current method enhancement. For efficient and safe transportation, the modernized physical layout and advanced communication infrastructure collaboration with sensing, computing, signaling, control and monitoring process are essential to support smart transportation.

1.2 HISTORY OF TRAIN CONTROL SYSTEM DEVELOPMENT

The primary goal of tracking each train is to ensure that they are operating in a safe and efficient mode. The general architecture of a typical train control system (TCS) is shown in Figure 1.1. It is considered as major block in the railway system, as it controls the movement, prevents collision between locomotives and regulates the service. The railway control system is followed by other blocks such as operation, signaling, data transmission and service control system.

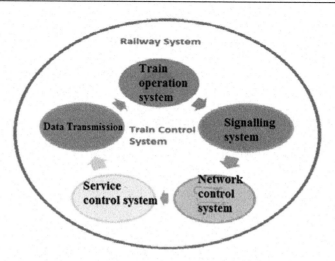

FIGURE 1.1   General architecture of train control system.

The evolution of railway signaling and communication involves basically four generations. In each generation, the incremental improvement in design, signal availability, communication channel and operational performance are major factors.

First Generation: It includes track circuit, wayside component and on-board component. Train-operating modes are restricted to manual driving modes. The train is detected by track circuit and wayside signals are useful in providing indication to drivers. The operational performance is decided by track circuit, fixed block configuration and wayside signal. In this generation, all control equipment is located on the wayside, with on-board train equipment limited to only stop indication.

Second Generation: It is also track circuit-based, but wayside signals are replaced by in-cab signals, which transmit speed codes to the train. The part of the control equipments is transferred to train for detecting and reacting to speed codes, and displaying movement information to driver. The train operating modes are permitted to manual driving modes, but operational flexibility is still limited by track circuit configuration and the number of available speed codes.

Third Generation: The evolution in the system continues the trend to supply more accurate control of train movement. The amount of data transmitted includes speed codes as well as distance, rather than responding to only a limited number of individual speed codes as explained in the second generation. This generation supports automatic driving modes, but movements are still decided by track circuit.

Fourth Generation: It is an advanced generation because of the improvement in efficiency and safety of train operation. It really supports automatic driving mode with a highly reliable and safe complex computer tool. It consists of four major parts: (1) central unit, (2) station and wayside unit, (3) onboard control system and (4) communication network. The generation includes advanced design, development and implementation for special line application, simulation field test and verification and safety measurement. Some of advanced TCS such as European Train Control System (ETCS), Chinese Train Control System (CTCS) for main line railway and Communication-Based Train Control System (CBTC) and Positive Train Control System (PTCS) for transit system are in use.

1.3 ADVANCED TRAIN CONTROL SYSTEM

The system is believed to be needed for smart transportation in rail sector. The increasing speed with environmental condition is the most preferred factor for train journey making it more and more difficult to operate safely without the assistance of technology. Advanced design and development require major technology correction to successfully balance performance, safety and reliability. The increasing complexity requires the best technical group to create solutions that match the depth of challenges required by rail industry. Different countries and organizations are developing their own train control system. For example, Americans developed PTCS and CBTC for their railroad. European countries implemented ETCS, while Chinese Railways created CTCS and Japanese introduced EJTC. They differ from each other in navigation, data transmission, and integrations of new components as well as simulation methodology. Figure 1.2 depicts the on-board and track control system.

In this section, we briefly discuss the major segments that are used in each tracking and controlling system. We also present additional information on how these technologies are helpful in solving real-world railway system problems.

1.3.1 Positive Train Control System

This rail safety system is designed to prevent train-to-train collision condition, derailment due to excessive speed and accident in work zone limit. It can identify the exact location and determine the speed. It will then automatically apply the brakes to achieve the desired movement. An end-to-end solution

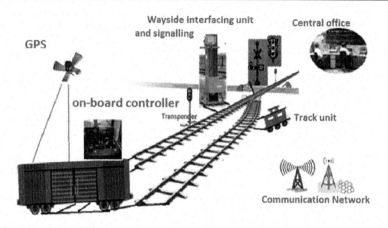

FIGURE 1.2   Scenario of different control system on track and train.

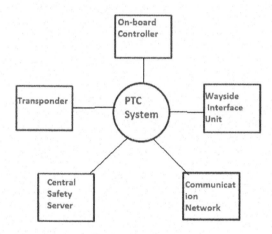

FIGURE 1.3   Positive train control system architecture.

consists of four major segments. Figure 1.3 depicts the architecture. We present here main architectural components separately summarized one by one.

On-board System: The onboard computer located in a train receives the information from wayside device and Central Office. The operator uses this data and takes appropriate action to limit speed and other safety concerns. If the operator does not slowdown or stop the train within 15 seconds, the onboard computer automatically applies a brake to stop train. This event must be completed before entering the next block.

Wayside Unit: It consists of signaling equipment on-and-around the track. The equipment includes track circuit, gate, lamp, switches and much more. The devices are connected to a wayside server through an interfacing unit. This sends information from trackside equipment to the Central Office for processing and to inform the locomotive directly and act on it accordingly.

Communication System: The communication system between locomotive, wayside device and central office depends on a bidirectional communication network. The PTC solution offers different communication interfaces such as Ethernet, 220 MHz PTC Radio, Wi-Fi and 4G Cellular. Here, onboard equipment uses 220 MHz radio or 4G cellular interface. But wayside equipment communicates with Central Office from any one of four-interface types.

Central Office: The function of the Central Office is to store, process, and act on information it receives from the on-board locomotive computer, and wayside messaging server. The database maintains information on tracks, trains, work zones, and speed restrictions. Based on this information, the Central Office sends movement information to locomotives.

1.3.2 Communication-Based Train Control System (CBTCS)

This is known as the most intelligent and integrated control system in rail system. It is adopted in many railways including mainline, light rail and underground line in cities. With the development in data communication, computer tool and control technique, CBTC represents as one of the best automated rail control systems. Figure 1.4 shows the general architecture of the CBTC system. At present, it has been used in underground line and light rail. It has not been implemented in mainline railway for many reasons. It also acts as brain and nerve center by providing safety and efficiency of rail system. The system is divided into five segments. The function of each segment is summarized as:

On-board Control System (OCS): The control unit is equipped with locomotive. The main function is to control train speed such as braking, acceleration, deceleration and cruising. In CBTC, onboard control system is most intelligent compared with other traditional systems. The train position and speed data information are sent to Block Control System.

Station Control System (SCS): The interlocking system is controlling the switch, signal and route at station. It communicates with Central

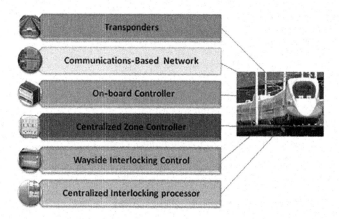

FIGURE 1.4 Communication-based train control system architecture.

unit, Block Control unit and On-board unit. Nowadays, most of the Station Control System is computer-based interlocking system.

Block Control System (BCS): It includes a radio block system and wayside equipment along the track. It always communicates with Central office, Station and On-board Control System. The main function is to control train operation in blocks.

Communication Network System (CNS): It connects other four systems as communication channel. It includes communication between train and wayside equipment in order to provide real-time, accurate, reliable and safe data exchange among them.

Central Control System (CCS): The heart of CBTC is the central control system. All the train operation is generated, dispatched and commanded in CCS with respect to train graphs. It receives the data from station, block and onboard control system and train operation.

It is obvious that CBTC will gradually replace the traditional train control system in railway network throughout the world. The traditional train control system is upgraded with the development of new technologies in the key technical areas, hence transforming into CBTC. The accurate position and speed measurement system are also the key technologies of CBTC systems.

1.3.3 European Train Control System (ETCS)

It is a subsystem of European Rail Traffic Management and also standardization of railway signaling system for European Railroads. The main goals of ETCS are summarized as:

Less On-board Equipment: There is a single on-board system and standardized Man-Machine Interface is provided.

Safety: The safety of train operation will improve by providing Automatic Train Protection (ATP).

Interoperability: The trains are interoperable across borders in different countries in Europe. It also requires reading the signals and considering as operator interoperability.

Cost-Effectiveness: It consists of only a few components. Hence, its manufacturing and maintenance costs could be decreased automatically.

Capacity: The rail line capacity is increased from 10% to 30% after ETCS application in railroad because it provides smoother operation.

Availability: The system offers less on-board, wayside equipment and less communication connection.

Open Market: There are monopolies for a railway signaling system in Europe to start of new ETCS project.

The ETCS monitors the movement of trains and other rail vehicles with local speed limit and maximum speed of the train. It also monitors route map, direction of travel and train operating procedures. It requires on-board and track elements, as well as communication system (GSM-R). The applications of ETCS are divided into several levels. They are:

Level 0: The on-board system called Automatic Train Protection (ATP) is installed in locomotives.

Level 1: It includes on-board equipment, balises or Euro-loops are added to the wayside equipment, and transmission channel is implemented.

Level 2: The unit includes on-board, radio system (GSM-R) is applied between train, wayside, and fixed block system is implemented.

Level 3: The system based on radio and moving block system is implemented.

In ETCS, balise is considered as very important device because it connects the communication between on-board and wayside systems. The track circuits are still required to detect train and send the position data to the Radio Base Centre. The data centre calculates the new data profile and communicates it to the operator. The computer is equipped with a train which calculates its location and speed data and sends it to the operator for the next braking point. Then the received message is passed on to the operator. During journey, the train will communicate with the balise as it obtains the modified position information. The unit fixed to the train is ready to determine the position and calculate

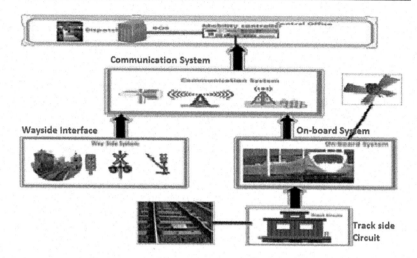

FIGURE 1.5   European train control system architecture.

the current speed for the distance traveled by train. The architecture of ETCS is displayed in Figure 1.5.

1.3.4 Chinese Train Control System (CTCS)

Based on the European system, Chinese Railway is also planning to drop the different signaling systems on their network. The Chinese Railway needs CTCS for high speed, and conventional lines, passenger and freight lines, respectively. Based on the present environment of signaling in the Chinese Railway Network, it is classified into several levels.

Level 0: It is the most basic train control model for CTCS. It is useful only for trains with a speed of less than 120 km/h. The system consists of existing track circuits, train operation system and cab signaling. The wayside signals are main signals, and the cab signals are supplementary signals.

Level 1: It is useful for the train with a speed between 120 and 160 km/h. It consists of track circuit, transponder and ATP system. The control mode for ATP is speed as well as distance. The block signals are removed and train operation is based on the on-board system. Balises are installed on track. The requirements for track circuit in blocks and at stations are higher than those required in level 0.

Level 2: This system is useful for trains with a speed higher than 160 km/h. It includes ATP, digital track circuit and transponder but no wayside signaling in block. The control mode for ATP is distance. A fixed block mode is still applied.

Level 3: The system with track circuit, ATP and transponders are major components in level 3. The track circuit is not transmitting train operation information but is only for occupation and train integrity checking. The train operation information is transmitted by GSM-R. A fixed block system is still applied.

Level 4: In this level, the moving block system is utilized clearly. Track circuits are used only at stations. GSM-R will transmit the data between the train and wayside components. The transponders, or Global Positioning System (GPS), are used for train position information. The on-board system is checking the train integrity data.

The general architecture for CTCS is given in Figure 1.6. The explanation of each control system is summarized below:

Track Circuits: It is always located on track line where the train passing information is communicated through wireless communication.

On-board Equipment: It monitors the safety operation of trains. It is used to generate dynamic velocity curves, distance graph and speed control modes on the basis of train movement.

Balise: It is used to transmit information about train location, temporary speed restriction to the on-board system.

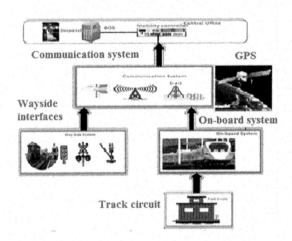

FIGURE 1.6   Chinese train control system (CTCS) architecture.

GSM-R: It is used as communication channel for bi-directional, continuous, more capacity of information communication between the on-board and wayside system.

1.4 INDIAN TRAIN TRACKING AND CONTROL SYSTEM SCENARIO

Since 1853, the Indian Railways has remained adaptive to new technologies in the field of communication and signaling. The introduction of relay interlocking, various block working, multiple aspects of signaling, train detection and radio communication technology, etc., was adopted by other railways across the globe. But Indian railways have adopted these technologies in a quiet and smooth manner in the past and succeeded in technology induction.

However, during the past 15 years, the import of advanced rail technology for Indian Railways has not been in the improvement stage. The LED-based color Light Signal lighting units, Integrated Power Supply System (IPS), Single and Multiple Sections Digital Axle Counters, Solid State Inter-locking, ETCS level II-based Train Protection & Warning System (TPWS), GPS-based Anti-Collision Device (ACD) and GSM (R)-based radio communication systems have been introduced.

1.4.1 Auxiliary Warning System (AWS)

It recognizes the failure of train operators and acknowledges the signal from the on-board system. The track circuit senses a danger signal and sends the information to on-board equipment. The brake system is activated as it is interconnected by the on-board system. The system will automatically apply brakes when any object appears on the track line, and train speed is reduced to a nominal value. Since 1990, AWS has been working progressively. Now, for a network of 330 km, AWS is working in Mumbai Suburban. Today, 2800 suburban train services are attached to AWS, run daily, carrying about 7 million passengers.

1.4.2 Anti Collision Device (ACD)

The device is GPS and Angular Deviation Count-based equipment to prevent collision of head on, side and rear end. It also provides a warning signal at level

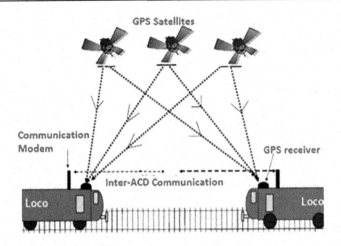

FIGURE 1.7   Architecture overview of anti collision device.

crossing. It consists of two types of ACD such as mobile and stationary ACDs. The stationary ACD is located at station and crossing gate whereas mobile ACD is equipped on train. They receive signals from the track circuit and communicate with each other within a 3 km range. When automatic operation is in failure mode, then the manual operation will compensate.

The architectural overview of ACD is presented in Figure 1.7. The model came into operation in 2006, covering about 1740 km of Northeast Railway. Advanced technology is needed to minimize unwanted braking, restarting, and safeguard for repeaters. But this system is not suitable for multiple lines.

1.4.3 Train Collision Avoidance System (TCAS)

The GPS-based ACD is facing a major problem in finding accurate track ID, which needs repeaters. On the other hand, the TCAS function is combined with ACD and ATP functions. It is designed to prevent train to train collision, speed restrictions and safeguard for train journey. The location information of the train comes from RFID tags located on the track. The unit calculates the speed with respect to movement authority, braking methods, target distance.

The stationary units will communicate the information to the remaining trains in that area via radio signal. From the information, it is concluded that whether the train operator has acknowledged the failure of brake in compensation with an automatic brake. The system uses a 19.2 kbps baud rate operating in the UHF range of communication for trains. The data is repeated automatically for all 2 seconds. The RFID tags calculate the location and also the distance travelled. Loco equipment senses the train speed through Tachometer. The braking system is interconnected with the locomotive unit for the application of brakes. The provision of an auto brake test is needed when a train senses a change in its configuration and there is no break test at SLVE.

To date, none of the new technology introduced listed above has been able to deliver the consistent performance in terms of reliability, availability and maintainability. Even after completion of their large-scale deployment, continued modifications brought to poor engineering performance. The advanced technology equipment application is finding more contribution on signaling systems which affects the safety of train operation.

The signal incidents are affecting the punctual running of train and not only causing dislocation of service and discomfort to passengers but also introducing the human element in train operation. In order to manage this growing traffic demand and maintain operational safety, new signaling equipment technology has been introduced to achieve higher reliability and availability of signaling system.

1.5 CHALLENGES IN CURRENT TRAIN TRACKING MODEL

The train tracking and control system explained above are automatic and intelligent systems. The signal indications are based on cab signals, automatic control and stop system. More specifically, they have limitations such as: comprised of too many components, depend on traditional track circuit, Infrastructure failure, traverse incorrectly lined switch, Wayside switch fails to detect position and fail to convey message, on-board equipment failure, insufficient braking force to de-accelerate the train, and missing speed limit in data base. Hence, the advanced technology is needed to improve rail service.

GPS is one advanced navigation technology that provides more precise location data. The faster detection rate is more safe than traditional techniques such as track circuits. Hence, the development of remote monitoring system based on GPS, Wireless Communication, Signal Processing and Embedded design with big data software computing are less expensive and less hardware component. For safe transportation, the GPS is combined with different communication network in order to reduce signal failure and accident due to collision, track damage information, train delay cause.

The future signaling and tracking system will be GPS based and on-board with a secure wireless data communication network that will be used to continuously monitor the location, speed and running direction of train accurately. It delivers robust performance under an adverse environmental condition of high communication jam and also noise. It is important that the future system will offer low cost and be able to adjust on any existing system to facilitate the replacement. The tracking model has also pitched the way for locating predefined place based on current moving object location mapped with other information loaded server to locate nearby services. Besides the real-time application of GPS-based tracking model, the reliability of tracking service is not clear as GPS-derived positioning information is dependent on surrounding condition as well.

Train Navigation Control and Information Management System

2

2.1 INTRODUCTION

Several challenging researches have been implemented for intelligent transportation system. In the rail sector, train tracking, controlling, monitoring and information management system play a key role in safety issues. On the other hand, this system provides an automated mechanism to enforce the safety of metropolitan transportation. In this field, developed countries like China, Japan, and European countries have already witnessed high-speed trains by adopting current technology platforms such as PTC and CBTC solutions.

DOI: 10.1201/9781003294016-2

2.2 ADVANCE TRAIN CONTROL SOLUTION

One of the most critical elements in signaling that assures the success of operating needs and benefits is to be realized by the advanced train control technology. It is also important to balance the needs and expectations of users with the capabilities and limitations of advanced train control technologies. Some of the operating needs are also recognized that any implementation strategy would need to accommodate the constraints such as improving the availability, reliability and accuracy, reducing rail cycle cost, flexibility for equipped and unequipped trains, supporting both manual and automatic train operation, and overspeed protection.

The comprehensive, integrated and intelligent control systems for rail systems with the development of modern data communication, computer and control techniques are established. In Europe ETCS, and in America, PTCS and CBTCS, in China CTCS, in Northern America and Japan, there are moving block systems or advanced train control systems. The discussion also supplements some of the advantages and limitations of various train control systems. The PTCS is designed on a signaling and communications technology platform to avoid accidents by operator or communication error. It is little difficult to limit the accidents by high speed, train movements, and wayside signal communication failure. But it does not prevent incidents due to trespassing on railroads level crossings as location, train movement and speed regulation can communicate through the sensors which are spread across rail track. If the locomotive is beyond the speed restriction or moving authority value, then onboard computer will display a slow or stop signal.

The original PTC testbed detection systems determine only the train's position to within a few feet. It follows the previous tradition of block-oriented detection and signaling systems. Trains operating on the new PTC testbed will be combined with a localization sensor package consisting of MEMS-based IMU. With Vital Electronic Train Management System and other real-time PTCS, the new testbed depicts precision navigation in the form of a sensor-based system carried by each train. This system will provide precise train localization information to the scheduling and safety units. A new model railroad testbed called UAH On-Track has been developed to explore the software safety engineering issues associated with PTC.

As mentioned above, PTCS is designed to avoiding collision or derailment. The government is reviewing the risk factors involved in the PTC deployment. However, from the regulatory, it approves the safety measure of PTC and its related issues. The query report is made assurance for the operating mode and environmental safety after the deployment of PTC. The advantages and disadvantages of PTC usage, operating mode, the risk factors involved, and rule requirements for analysis are expressed. The author also explains the critical failure modes of PTC and future work that would facilitate the risk assessment process.

When we talk about the more expansive version of the PTC system, it is the CBTC system. This would bring additional safety benefits in terms of accident avoidance. The train monitoring and controlling are integrated into a single system through data links between trains, central office computers, and wayside computers. It is one of the novel signaling and controlling platforms tailored for metro. It provides continuous automatic train protection as well as improved performance, system availability and operational flexibility of the train.

The basic architecture of PTC and CBTC systems consists of on-board models, wayside equipment, communication infrastructure and central office. For continuous monitoring of trains, it is required to have communication between wayside-to-train, train to central office and train-to-wayside data communication. The communication media selected for CBTC includes fiber optic cable, cellular network, radio system and GPS. The communication segment provides a communication path between office, track element, train and roadway. The above-mentioned communication system is applicable only for metro trains. When it comes to general, the entire communication as well as control system will change according to satellite visible area and low satellite visible area.

The automated and intelligent system includes too many components. The list of components are on-board components: Computer display units, locomotive messaging systems, GPS sensors, crash-hardened memory modules, network devices for communications. Wayside and switching network: integrated and stand-alone wayside interface units, communication switching network, switch position monitors; communication network for base station, locomotive and switch and signal communication Central office: back office server, train management computer, interoperable electronic train management system software, user authentication systems, track database, interface and future version to the delivery unit, data security-related application for message transformation and interconnected train controlling unit.

To increase safety solutions in railways, many countries are installing train control systems (TCS). Let us discuss the installation of such system scenario in China and European countries. In China, mostly the trains were operated by drivers manually with the help of warning signal from trackside circuits, which are located on the track mainline. It also includes a trackside interlocking-blocking signal, combined with other train signaling. Recently, high-speed bullet trains are in developed condition, and the traditional Track-Based Train Control (TBTC) has been completely replaced by CTCS. It has the specialty that it is deployed for a high-speed interoperability system. The CTCS is an ATP system based on cab-signaling and monitoring the train system data transmission. The Movement Authority (MA) and corresponding rail-line data are passing to the control unit of train. The data are displayed in on-board system for driver assistance. Based on the signaling system of Chinese railway, the CTCS-equipped train will operate for any Chinese railroad line with more functionality and not much technical issues, but the levels of CTCS depend on network signaling. The levels depend on mixed operation for interoperability where the line is equipped with a wireless system.

The ETCS monitors the movement of train with the maximum authorized speed limit of train concerned, adherence to the route map approved for train, direction of travel, operating procedures, etc. This requires on-board equipment, various elements along the rail-line and radio system (GSM-R). The specific interests of national railway sectors and different types of trackside equipment used in traditional train control and protection systems, configuration levels, etc. ETCS level is a system configuration that specifies the required equipment components such as balises installed in the track, radio block center and line side electronic unit. However, the levels are downward compatible, not upward compatible.

The ETCS was developed for the European Rail community with an ATP system by replacing the existing one. The Euro-balise is a local transponder providing trains with information on the route attributes and speed limits. It could replace balises used today by ATP systems on conventional lines. The system makes it possible to implement continuous speed control. ETCS is thus able to provide safety levels that are higher than ATP systems currently in service through Europe. On high-speed lines, the ATP systems are coupled with cab-signaling. The cab-signaling is also called Euro-cab. A radio system called as Euro-radio is GSM-R layer. It transmits the signaling information from ground to Euro-cab on high-speed lines. The main advantage of using radio transmission is to transfer data in both directions without installing equipment in the tracks.

Railway systems have a long history of train protection and control to reduce the risk of train accidents. Many TCS include automated communication between train and trackside equipment. However, several different national systems are still facing cross-border rail traffic. Today, trains for cross-border traffic need to be equipped with TCS which are installed on the tracks. To estimate the impact of GSM-R communication on the line capacity, it introduced a stochastic communication model to test track and train behavior. Research has been conducted simulations of railway system with simple track layout by means of two modeling approaches for comparison. The results of stochastic new models are comparable to those obtained by applying a sophisticated modeling tool. The simulation of ETCS level 3 indicates that a GSM-R communication failure does not have any severe impact on the capacity of rail-line.

Indian Railways has recognized requirements of accident prevention system through new technologies and signaling and telecommunication. Under the safety mission with zero tolerance for accidents, Indian Railways has introduced modern safety devices such as Block Proving Axle Counters (BPAC), GPS-based Fog Safe Device, AWS, and Vigilance Control Device (VCD). One such important advanced technological up-gradation to achieve reduced human dependence, Konkan Railway Corporation developed ACD also called "Raksha Kavach". It is a train collision prevention system. In December 1999, GPS-based prototype was demonstrated by KRCL. These are located on train, brake system and at stations and gates of level crossing. These communicate within the network and helping in avoiding accident. The types of ACD are, mobile for locomotives and brake vans, and stationary for stations and level crossing gates information.

2.3 GPS AND DIFFERENTIAL GPS-BASED TRACKING SYSTEM

The traditional TCS based on track circuits for detection information requires deployment and management device. Numbers of reasons are quoted as to why conventional systems show less grounded results that include: ground facilities that installed on and around the track; more signal cables are required; train positions cannot be detected accurately. With GPS as the tracking system and the communication between the on-board equipment, the advanced TCS is designed and calculates the train position information.

In Satellite Visible Environment (SVE), tracking the train using one of the best satellite navigation systems called GPS. But the service failure occurred in Satellite Low Visible Environment (SLVE). The satellite-based navigation system is used in SLVE to calculate the train journey information. In black territory environment such as tunnel, bridges, valleys and many more, the location-based information is very weak. This is due to there is no line of sight between transmitter and receiver. The satellite-based navigation method is seeming to be more accurate, efficient, and less economic maintenance.

The loss due to multipath in the SLVE will reduce the information completely. Generally, it is believed that the navigation information cannot be completely dependent on this system alone. Laser scanning, aerial vehicle surveillance, Radar and Sonar are some of the expensive surveying techniques but restricted with area and time can also be used for geographical data.

Even today, there is destructive terrain displacement; threatening critical socio-economic platform is increasing in SLVE. To enable modeling of the complex relationships between geological properties and ground stability information about spatial-temporal variation of terrain and surrounding microclimate is needed. Traditionally the geological data is acquired by means of manual or expensive surveying technique such as laser scanning, interferometry synthetic aperture radar and unmanned aerial vehicle surveillance.

In addition to high cost, these cover only very restricted areas and short periods of time. The lack of spatial and temporal resolution severely limits the analysis of large-scale terrain dynamics and their origin. Hence, the availability of GPS technology and algorithmic advances has enabled every researcher to use single frequency L1 GPS receivers for terrestrial and aerial positioning applications. The inherent errors in GPS positioning caused by atmospheric effects, multipath interference and clock uncertainty are mitigated with expensive, high accuracy dual frequency L2 GPS receivers which exploit two distinct GPS satellite signals. Hence, differential GPS techniques with exploitation of three distinct GPS satellite signals using the Tri-lateration technique to achieve more accuracy in detecting the moving object. A navigation process avoids line of sight error, which improves GPS localization.

Many researchers have done a lot of work on train tracking, with measurement of speed and velocity in SVE using satellite-based navigation system called GPS. It is one of the advanced technologies which are widely used for tracking, positioning, surveying and navigation because of less maintenance demand. However, the authors came to the conclusion that GPS technology is not suitable for SLVE such as forest, bridges, under pass, tunnel, and many more. In real-time application, the moving object information is reduced because of no line of sight between transmitter and receiver and loss due to multipath. The use of differential GPS is proposed to identify the specific location of the train which improves the accuracy. Generally, DGPS operated with two types of receivers. The signal computing timing diagram is done with roving receivers and stationary receivers which calculate satellite position. Differential GPS use many stationary and mobile-type receivers to increase measurement accuracy. In coordination with base station, the mobile receivers are calculating their absolute positions with increased accuracy.

2.4 WIRELESS SENSOR-BASED TRACKING SYSTEM

The history behind wireless technology enables everyone to focus more on the sensor concept. The satellite-based navigation with wireless sensor technology can make a big boom in communication. In the measurement of environment conditions such as temperature, pressure, humidity, surveillance of moving object in water, air and water and also pollution monitoring, etc. where the wireless sensors are used. The data collection from object is done by wireless sensor network technology because of its sensing, identifying, monitoring and controlling the object in a pre-defined path.

Next, the integration systems utilized in the implementation of an intelligent train monitoring system using different sensors such as RFID, GPS, GPRS (Global Positioning and Radio System) and GIS (Geographic Information System). A new theoretical frame and rule-based decision algorithms are developed for prototype implementation. The integration of RFID technology and clustered WSN in order to track vehicles is an alternative to GPS-based system. High cost in installation, maintenance, limited coverage problems in GPS-based measurement.

One more methodology is needed that indicates the localization and communication of train in satellite visible and inside the tunnel are to be measured with great precision. In such cases an integrated system that helps in radio communication system and location solution with low budget railway. The implementation of the GPS-based train monitoring system called GPS-GPRS module transmits the location information to web server. It is believed that the GPS-WIFI, and RFID-pedometer system is proposed with the integration of sensor for complex environments. This system will supply higher precision output and solve the advantages and limitation in each localization technique. For accurate navigation then increase the movement speeds of moving object. For advanced navigation in SVE and precision identification near object, WSN-RFID based model is the best choice.

Some researchers have proposed the Zig-bee-based WSN localization method because of low economic maintenance and less power consumption. Extended Kalman Filter approach is used for analysis in a co-operative manner for WSN and Zig-Bee data information. The EKF is used to regulate the filtering process and convergence of the referred scheme. The concept also investigates how the line of sight and loss due to multipath will decrease the accuracy of GPS in real-time navigation application.

2.5 DESIGNING OF TRACKING CONTROL MODEL AND DATA INFORMATION ALGORITHMS

WSN is widely adopted in object tracking application because of the advancement in data quality and monitoring accuracy. But WSN is facing some of the issues such as noise- interference, bandwidth restriction, non-linearity and many more. Hence it is a challenging area to study detection and monitoring-based problems. The object tracking and control issues in the WSN environment are presented with advance control technology. The controller was designed based on a fuzzy observer model for a nonlinear system with moving object range and rate determinant for tracking with data information collection.

The navigation system can also be formed by the integration of DGPS, three accelerometers, gyroscope along with barometer. The barometer is used to measure altitude pressure, which in turn estimates the change of road grade, which is needed to remove the gravity component sensed by accelerometers. The accurate measurement requires information about pressure and temperature at mean sea level. The system is most suited for hillside navigation with reduced sensor outputs during GPS outages of less than 1 minute.

The self-contained dynamic-aided error correction method is useful to overcome rapid navigation error drift generated in the absence of aiding sensors. It is observed that the dynamics dependent variables are measured during stationary and straight-line motion of vehicle. The vehicle dynamic identification system used is fuzzy logic. It is reasoned out that this method is highly suitable for land vehicle navigation in urban canyon, where severe GPS signal degradation, frequent vehicle halt and turning dynamics exist. A novel adaptive fault-tolerant multi-sensor navigation strategy for unmanned vehicles on the automated highway system is also based on DGPS. It explains about the generation of reference trajectory path by means of INS, up-gradation of position and orientation by means of CP-DGPS, Wheel Encoder and Electronic Compass. It is observed that in this method the fuzzy logic has also been implemented to obtain a highly reliable navigation system that can assure safe running of automated vehicles in complex situations. Even in cases of sensor failures, system performance does not deteriorate.

The trade-off between accurate and fast localization and tracking algorithms has very high computation and memory requirements. Constrained devices such as wireless sensors usually require algorithms with low computational and memory requirements. The Kalman Filter (KF) has initially been proposed for linear systems. It comparatively requires low computation power compared to other Bayesian algorithms and has been extensively studied in the past few years for its usability in wireless sensors. The main limitation of KF is requirement of linear state information and estimation equations. The EKF algorithm is most widely used for nonlinear systems in linearization of the observation model. The EKF has been extensively used in many areas including probabilistic robotics, neural networks, image processing or depth recovery and Global Navigation Satellite Systems. The EKF is an extended version of KF to better handle non-linearity. EKF core basically remains the same; however, its computation complexity may vary depending on how state information is described or presented and updated. The conventional EKF and IEKF algorithms do not take into account the linearization errors and lead toward inconsistent state estimates.

The real-time implementation of KF algorithm made big history for tracking while the vehicle is operating. The experiment is conducted over the same measurement data that could be fed to particle filter with large number of particles. The performance of KF compared against particle filter as validation of how accurate it is. It uses a common algorithm for a sequential inference that provides optimal estimates for the current state and covariance given a linear system with Gaussian noise.

The two analytical methods called Tri-lateration and Triangulation are used for tracking and navigation application. The Tri-lateration method use only distances measurement to identify the position of objects whereas Triangulation use angles and distance to locate an object. The application of these two-analytical tracking and navigation methods to a dynamic system has some disadvantages like distance measurements become very fluctuating and noisy which in turn makes localization becomes more difficult. This requires a suitable filter to remove the unwanted noise signal for better accuracy. The Quadratic Optimal Control (QOC) theory is adopted in data fusion model to check the tracking control performance. There are many solutions for QOC problem. But with the Liapunov approach, the direct relationship between Liapunov functions and quadratic performance index can be performed and by varying performance index value with respect to control inputs, can measure data losses in the operation.

2.6 TRACKING CONTROL MODEL USING MULTI-SENSOR DATA FUSION

The data fusion technology is one of the curve edge technologies in many summing data application, where more inputs are combined, processed and rectify to obtain high-quality outcome. When the data from multiple sources are combined together to provide minimized and full description of the process. The data mixing technology is applicable in many real-world applications such as in defense service, surveillance of land, air and water monitoring system, and in automation, etc.

The maneuvering and non-maneuvering object monitoring is a hot topic in a smart air-based application. Some of the major fields where sensor techniques are mainly used is image processing, automation, etc. The tracking efficiency can be improved with the fusion method. The monitoring of objects using WSN is a major milestone in sensor technology. Sensors are mainly used in real world such as robotics automation, military-non-military and wireless systems. To overcome the defects of the current statistical model on non-maneuvering target tracking, a novel fusion algorithm for tracking large-scale maneuvering target is adopted. The fuzzy adaptive Kalman filtering algorithm with maneuvering detection was used for a large-scale maneuvering target. Later data are extracted from Kalman filtering processes to estimate the magnitude and time of maneuvering. The tracking system with both active and passive radar has higher precision than those with single sensor for large-scale problems.

Technical survey highlighted the data fusion technology and designing of models based on state and observation process. Bayes theorem is considered as an excellent algorithm which connected with the data mixing process. For sensor modeling, the probabilistic and information-theoretic methods are used. A probability theory deal with uncertainty manipulating set of methodologies explains architectures, fusing methods and maintains data information. A multi-sensor-based project highlights the value of Bayes theorem in producing observed data information with prior observation of probabilistic matrix state. Number of sources has advantage in providing a direct format of data interpretation of observation with lots of information exchange. Based on the current result of data fusion tracking algorithm, a novel technology using the Bayes theorem is preferred most. Implementation by this theorem will suppress the irrelevant data and noise in fused results to a higher level of precision and stability.

The scenarios aimed at demonstration of major key elements of advanced filtering process. Filtering in modified processing of measurements values, computation of coefficients smoothing, accumulation of results is taking place. The Information filter is advanced version of discrete-time KF. The state estimates and estimation covariance in KF are replaced by information matrices and information vector. To deal with real-time source data fusion problems, information filter is one of the best filter methods. The filter has excellent special qualities in exchanges of data information and interpretation sources observation directly. Based on the current result of data fusion tracking algorithm, the author presents novel technology using Square Root Information

Filter Algorithm (SRIFA) for possible application to data fusion process. The Information filter has a sensor network in a decentralized pattern. In a decentralized sensor network, nodes are exchanging data within the network itself.

On the basis of centralized data fusion algorithms, data fusion decentralized architecture is constructed. While comparing all common data fusion algorithms, the decentralized algorithms are more feasible in terms of computation complexity and communication criteria. The initial presentation for decentralized systems is the development of decentralized form of KF algorithm. This is achieved by placing the original KF state estimation problem in the information form. The summary of KF algorithm, data fusion concepts, and basic KF algorithm is assumed and corresponds to four major algorithms such as group-sensor, sequential sensor, inverse covariance, and track-to-track fusion method.

Developing the theoretic framework on multi-sensor data fusion is taking context into consideration. The method is to combine numerical and symbolic information, in order to have fusion process in different levels. The level of treatment that analyzes the context data using contextual variables for the estimation process model is designed. The outcome result is dominating the measurements provided by sensors well-adapted to maximum context data and to minimize the not well-adapted sensors.

With the LQR method, the controlled object is analyzed by optimization is done for tracking target. In every case, the network traffic congestion is eliminated by the energy factor bandwidth. The major purpose of the research is to realize the control theme of network such as QOP and QOS and optimization of network in NCS code design. The control theme of network is improved when bandwidth is considered for checking control performance and stability of the network. Finally, the bandwidth control theory is employed to realize network scheduling. The feedback controller design is mainly depending on the following factors. They are control plant model and the state estimation information. The model for statically changing parameter of LTI system with interconnected optimal state-feedback controllers. The linear system parameters are considered as random variables which are distributed identically. By designing optimal state-feedback controller such that it acts as a liner in the finite state and nonlinear in infinite region based on traveling path and model parameters.

Generally, for the LTI system, the state-feedback control design is explained based on H_2 and H_∞ guaranteed performance is addressed. It is useful specifically when the control law design with state-feedback gains and its switching function. A scalar variable where the simple alternative conditions are based on Lyapunov or Riccati-Metzler equation with fixed scalar variable, the conditions are expressed as LMIs. In order to increase the system performance, communication channel sampling periods are deal with self-triggered control design problem. The Continuous-time Differential Lyapunov Equation (CDLE)-based analysis is adopted for estimating and filter of state values of continuous-time data samples. By comparing these stochastic process values with Ordinary Differential Equation (ODE) of discrete data. The methods presented here are initially discrete the system and next apply Discrete-time Difference Lyapunov Equation (DDLE), over sampling is used to manage stability and accuracy.

The self-triggered control problem is considered as most advanced technology because of less communication network and closed-loop task. With the execution of resources without changing their communication channel and feedback loop behavior than a traditional periodic time-triggered algorithm with classical LQR self-triggered problem. Strategy is depending on levels of performance which are explained in a quadratic materialistic pattern. The LQR problem is a special case of self-triggered methodology where specifications are considered as (1) triggering mechanisms designed with pre-defined control rule; (2) methodology and communication requirements and (3) feedback designing is proposed with control law.

The designing complexity in the control system mostly appears to be in a nonlinear dynamic system. It is a challenging area where unmanned aerial vehicle actuated design configuration becomes a unique platform for the research community. There are a number of algorithms that makes analysis for certain theoretical explanation of the system are PID, fuzzy logic, artificial neural network, linear quadratic regulator, etc. Several other optimal algorithms have been analyzed depending on their advantages and limitations for integrated systems.

Hybrid System for Train Tracking and Monitoring Model

3

3.1 GPS-BASED TRAIN TRACKING SOLUTION

The GPS utilizing system of many satellites was invented for purpose of navigation but developed and maintained by the US Department of Defense (DoD). It became fully operational globally in 1993. A total of 24 operational satellites were used initially, distributed 12 over six orbits in properly geometrically spaced slots. The satellites orbit around the earth about 20,200 km above the earth's surface with approximately 12-hour periods. Each satellite transmits a radio signal containing navigation parameter to estimate its position, velocity and time information. GPS signals mainly consist of radiofrequency carrier, unique Pseudo-Random Noise code (PRN) and binary navigation message. The PRN code is a sequence of 1's and 0's multiplied with binary navigation message and modulated with carrier frequency 1605.32 MHz and 1328.40 MHz to form the transmitted signal. The GPS receiver acquires the transmitted signal information and performs data processing.

To estimate Time of Signal (ToS) transmission, it is required to calculate satellite position, velocity and time information. The ToS transmission combined

DOI: 10.1201/9781003294016-3

with time of reception using an on-board GPS clock. This in turn is used to estimate Time of Arrival (TOA) of signal. The satellite to user range measurements are calculated when TOA information is multiplied with signal speed and velocity of light. If the GPS receiver clock is not perfectly synchronized with satellite clock, result in biased TOA measurement and thus produce biased range measurement known as pseudo-range. A navigation system such as GPS delivers sufficient long-term accuracy to millions of users on the earth. This has dramatically increased GPS real-time application in various diverse fields such as land vehicle navigation, earthquake monitoring, bridge monitoring, fleet management, traffic enforcement, surveying and in critical application such as emergency service vehicle (Ambulance) precise navigation to the accident zone.

GPS-based navigation technologies are useful in navigation of land vehicle. The position and velocity information are received from four satellites. GPS-based positioning has a significant way for locating the place of interest based on current user location matched with other information which server to locate nearby service. The GPS-based navigation information accuracy is less reliable and depends on surrounding environmental-related conditions. In other words, system localization accuracy is limited when it passes through SLVE such as bridges, tunnels, underpasses and many more areas due to weak signal strength and distortion. Thus, to improve system accuracy and reliability, it must be integrated with other wireless sensors capable of bridging the gap.

For the last few decades, many researchers have done a lot of work on train tracking system by measuring speed and velocity in satellite visible environment using GPS. It is one of the advanced technologies widely used for tracking, positioning, surveying and navigation application. Even though GPS is the most advanced technique for navigation, it delivers the required accuracy only under certain conditions. Although GPS devices become more and more available, there are still some situations where they cannot be used.

1. GPS signals are sensitive to obstacle, which make indoor positioning hard to implement.
2. The accuracy, precision, price and size of GPS receiver are prohibitive for many real-time applications.

3.2 DGPS-BASED TRAIN TRACKING SOLUTION

The aim of the current research is to demonstrate a method to track train using automated technologies such as GPS and report their location automatically. But knowing train location through GPS satellite may transmit information data on a certain part of track with great circulation delay. Hence, depending on the accuracy of positioning service of the system, sending surveillance information from train is easy for analysis.

In this section, we present some of the positioning services available at the real-time application level.

1. **Precise Positioning Service (PPS):** The military and United States of America users will provide with maximum accuracy.
2. **Standard Positioning Service (SPS):** The civil and other users will offer with less accurate positioning than PPS. It provides horizontal positioning accuracy of about 25 m with 95% probability.

Scenarios of the type of GPS which are well suitable for object tracking solution are:

1. **Assisted Global Positioning System (AGPS)** provide within less than 50 ft when the user is in satellite visible environment and within 160 ft when they are inside the building.
2. **Differential Global Positioning System (DGPS)** provide within less than 3 ft availability and employs both roving receivers that make satellite position measurement and stationary receiver that use their position to compute signal timing.

The most advanced version or enhancement to GPS is DGPS. It provides more precise detection accuracy than other navigation systems. The detection of moving object increases when they are coordinated with the GPS receiver. DGPS results in two differential corrections to GPS values in order to maximize detection and monitor the radio transmission of GPS satellites. With a DGPS receiver, the kinematics update of any moving object will be better than 2 m.

3.2.1 Selection Factor for DGPS-Based Tracking Solution

The selection criteria for the satellite-based tracking system adopted in this research depends on the level of service standard of DGPS and tracking accuracy level. Following are some of the listed parameters that affect the working model.

- Signal availability is 99% of the time.
- Accuracy to 2 m. 98% of the time.
- Integrity monitoring warning system within 10 seconds.
- Broadcast reliability 99.8% of the time.

- Enhanced coverage in all zones.
- In the ionosphere, tracking errors are removed half up to 0–30 m.
- In the troposphere, tracking error completely removed up to 0–30 m.
- Signal noise up to 0–10 m completely removed.
- Ephemeris data up to 1–5 m are completely removed.
- Clock drift up to 0–1.5 m is removed.
- Multipath up to 0–1 m are not removed.

Figure 3.1 shows the good precise tracking of train when numbers of satellites are placed widely.

In addition, there are a number of factors that will affect the accuracy performance of a differential GPS-based tracking system. Those include:

- Corrections with distance from differential monitor station.
- Update broadcast rate of differential correction.
- Selection of area for installation of differential monitors station.
- Variation in geometry consideration of satellite constellation.
- Form of the intentional degradation in accuracy method employed.
- Type and method of differential correction.
- Direction of the user-monitor end.
- Tracking algorithms, Ionosphere properties, earth's curvature, etc. also affect accuracy performance.

We know that GPS measures the position of the target on earth by receiving signal sent from four satellites. This system works with reduction in accuracy due to some errors. As locating the object, GPS uses radio signals which travel through the atmosphere at the speed of light. But earth atmosphere slows down

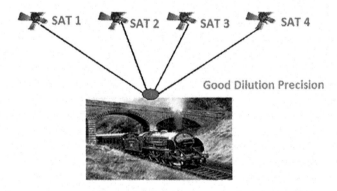

FIGURE 3.1 Relationship between satellite position and precision.

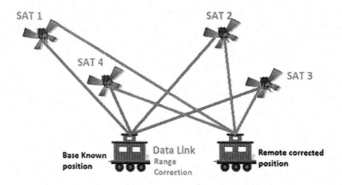

FIGURE 3.2   Working principles of DGPS.

electromagnetic energy while traveling through ionosphere and troposphere. DGPS will correct the errors when there is a misconception between locations and atmosphere. In differential GPS process, stationary station is known as reference station. This station calculates differential error and makes differential correction for location and time. This stationary station after corrections broadcast these radio signals to all DGPS equipped receiver which give an accurate location of object which are much more accurate than ordinary receivers.

In the 21st century, the tracking enhancement offered by DGPS makes milestone on the navigation system. The smart identification, information from E-source and map display system are mainly used for location information. Figure 3.2 shows the working principle of DGPS.

To overcome the shortcomings of GPS, Differential GPS is the better choice for tracking train. Unlike GPS, Differential GPS is capable of providing high dynamic information with excellent tracking performance. Generally, DGPS is operated based on two receivers namely mobile and stationary receivers. Where the satellite position is calculated by mobile receiver and measured position signal to timing signal conversion is done by stationary receiver in coordination with a base station.

3.3 INTELLIGENT RAILWAY SAFETY SYSTEM WITH RADIO FREQUENCY IDENTIFICATION

In the electronics field, communication system and information technology, a major advancement boom in machine-to-machine and object-to-object communication technology creates the sprawling effect. The increasing expectations for systems are cheap, efficient, less power consumption, excellent security and safety measures. Nowadays, RFID technology has moved from bar-code application into object identification applications. When we talk about automation, RFID is considered one of the most effective automatic identification technologies for an intelligent vehicle system. It mainly comprises components such as reader, tag and host computer. The data such as product source, manufacturer data, etc are stored by tag. It is attached to the targeted object. If the tags are operated with read and write then it is called active and it is called passive with no power source. The reader transmits the read operation signal to passive tag. The active tag is capable of maintaining two-way communication which reflects transmission signal to higher domain with communication range up to 100–200 m.

There is no regular communication between tag, hence the communication is single hop. For collection and processing of data, the reader reads information on the tags and transmits the message to host computer. The reader transmits all data to all tags within the pre-defined range.

3.3.1 Advantages and Application of RFID

The following are some of the advantages and also suggestions to use RFID.

- Tags are not affected by any chemical reaction or dust, etc.
- Tag detection does not require human intervention. Hence, it reduces employment cost and eliminates human error.
- Tag can be placed anywhere because it does not require Line-of-sight.
- Ability to identify object individually rather than generically.
- Tag can act as a unique identifier with large amounts of data storage capacity.
- When tags are combined with sensors then data can be read simultaneously.
- It reduces time lags and inaccuracies in data collection because the automatic reading takes place at several points simultaneously.

The major sector where RFID is extensively used mainly in tracking and management of retail stock, library books, vehicle parking slot, security key, automatic toll collection, healthcare and many more. The following are some of the main real-time applications where RFID can fit into it.

Instance or Class Identification Data: The normal database is required when RFID tags are used for product identification. The support system determines the location of product or processed handled is determined for a particular product.

Location Identification: The location of the object information is read by reader. The function is to track the present location of a given identifiable object.

Transfer of Further Data: The data from the tag are in the form of user unreadable format. Hence the impractical or impossible remote or pre-recorded database is applied for processing. The information about the processing of product can be written on the tag as data.

Asset Tracking: It is hard to locate where tags on assets are lost or stolen. It chose a real-time locating system that uses an active RFID beacon to locate container within 10 ft.

Manufacturing: It is used in manufacturing plant to track part, check work in process, to reduce defect, increase efficiency and different processing method for the same product.

Supply Chain Management: This type of technology has been used in automatic part, closed-loop supply chain within the prescribed system. It is also used for tracking shipment among supply chain partners.

Retailing: The retailers are using advanced techniques to improve stock security and management of product making.

Payment Systems: The real-time applications of RFID is automatic road toll collection. The same quick service is utilized in some restaurants, are experimenting with RFID tag to pay for meal at a drive-through window.

Security and Access Control: Used as an individual swiping card to monitor any one who has access to the office building. Previously the access control systems were used low-frequency RFID tag. Recently, RFID has been introduced 13.56 MHz system that offers longer read range. The advantage of having RFID in office is that an employee can unlock door using a badge rather than key or swiping magnetic stripe card.

Railway System Using RFID: The technology application in railway system help to ensure safety enhanced operation and boost the service level in rail transport industry. Using RFID, real-time applications are set up to provide the following benefits:

- **Operation Safety:** To determine train location, monitor and control the positioning of train to stop at a defined position.
- **Rail Asset Management:** Railcar, track equipment, level cross-ing signal equipment and infrastructure to ensure maximized asset utilization.
- **Service and Customer Satisfaction:** Receive real-time informa-tion on train location and update passenger information display at station and terminal.

3.3.2 Train Detection and Data Exchanging System

In real world, locations of trains are communicated to signaling Center by track circuits or axle counter. But one such technology that is in need to identify, detect and monitor train in all possible environments is RFID solution. It consists of tag and reader. An electronic tag carries a unique identity and is mounted on a moving train. Railway reader units are installed in several locations (Nearest stations) throughout the railway to enable the reading of RFID tag information.

When the reader receives the signal from the tag, it sends to central computer via GSM. Tracking and monitoring of all information is possible in real time through server. The information collected by computer is used for a number of purposes including producing an informative display of which train has just arrived at which location. The detail of a particular train such as train number and number of wagons also continuously monitored. The tag data are automatically configured, operate and monitored. A number of commands of configurations are predefined to reader which specify when, how and what to read and do with tag data. In this mode once configured reader will start to operate on its own. Then host computer is set up to operate for notification message from reader with tag data. Figure 3.3 depicts RFID based train detection system.

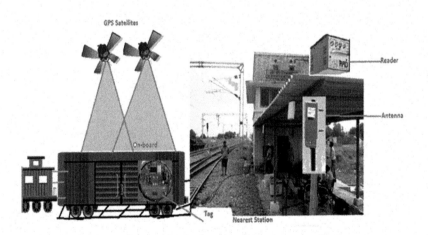

FIGURE 3.3 RFID-based train detection system.

3.4 FUTURE WITH WIRELESS SENSOR NETWORK

Wireless sensor network (WSN) has core attention worldwide in recent years. The sensors are smart tiny, inexpensive, with restricted processing and computing resource. These are attached with type of processor, memory storage device, energy supply, and an actuator. The nodes have interaction, calculate and collect information, then process it and finally transmit sensed data to user. To measure the environmental properties, numbers of optical, magnetic, mechanical, thermal, biological and chemical sensors are built to node. In some cases, the nodes are located in remote places; then the information is not possible to transmit because of less memory. The data information is transferred to the base station through wireless communication. Depending on the type and application of sensor used, actuator may be equipped in sensor.

WSN normally consists of sensor nodes working together in monitor areas to obtain data from the environment. There are two types of WSN:

Structured WSN: In this WSN, sensor nodes are arranged in a preplanned network. Some nodes are arranged at a specific location to provide wide coverage with lower cost for network management. Other types of sensor nodes are arranged in ad-hoc nature have uncovered region.

Unstructured WSN: It contains a major collection of nodes that are arranged in an ad-hoc fashion. Once the arrangement is made, then the network is ready to perform monitoring and reporting function.

The main great potential for application of WSN in real-time scenarios is listed below.

- In military, WSN is utilized for target tracking and surveillance as detection and identification.
- In disaster management, the nodes are useful in determining natural environment hazardous condition and weather disaster in advance.
- In biomedical applications, sensors help to monitor the health condition of patient.
- To detect the earthquake and eruption possibilities, nodes are arranged in ad-hoc fashion in the volcanic area.

The area of network, arrangement fashion and network topology are depending on environment. The area of a network depends on the kind of monitoring environment. For indoor environment application, less nodes are forming network while for outdoor environment application require more numbers of nodes to cover a larger area. The network connectivity is affected when there is an obstruction in environment also limits communication between nodes. Unlike traditional network, WSN has its self-structure design and properties availability constraint. The constraint includes energy utilization, communication coverage, bandwidth spectrum, processing analysis and storage capacity in each node. The design parameters are considered to be real-time usage oriented and depend on the type of monitored environment.

The sensor technology platform provides a solution in design and developing many kinds of wireless real-time applications. The sensor node is attached to devices that can measure environmental parameters such as light, temperature, humidity, pressure and many more. The function of the sensor node is to measure the reading from the monitored environment. Later, the data acquired from a generic sensor are passed to the base station for further processing and analysis because the nodes have higher processing, power, and transmission management.

3.4.1 Real-Time Application Based WSN

Depending on the application and environmental condition, the sensor network can be arranged on land, air and water. Depending on different challenges and constraints with respect to condition of the region, the following are classifications of WSN.

> **Terrestrial WSN:** The sensor nodes are arranged in a pre-planned or Adhvoc fashion. In an ad-hoc pattern, nodes are spread from high level and individually located in the pre-defined region. In preliminary pattern, sensor network is considered as optimal and grid placement models. Terrestrial sensor nodes are able to effectively communicate data to base station. The major characteristics of terrestrial WSN are, the nodes equipped with secondary power source, round off data redundancy, good transmission range, and multi-path routing with energy consumption, in-network data processing, minimizing delay, and using low duty-cycle operation.
>
> **Underwater WSN:** The typical underwater wireless communications are established from acoustic wave. The constraints are communication is acoustic, less requirement of bandwidth, longer data delay

and signal shadowing condition. Hence, for autonomous underwater vehicles exploration, these types of sensor nodes are distributed in water. The main characteristics of Underwater WSN are, the sensor nodes are able to configure themselves in ocean environment, more expensive, equipped with limited battery storage.

Underground WSN: It makes communication is very challenging due to data loss and high level of obstruction. The major characteristics are that it requires larger bandwidth, storage capacity of energy and maintenance cost consideration. It consists of a number of sensor nodes located in tunnel or mountains and under the ground are used to monitor underground condition while the nodes located above ground is sink node. It supplies information to base station later. It is more expensive WSN because of equipment, arrangement pattern, and cost. To ensure that communication through soil, rock, water, and other mineral content is more effective hence equipment required for this process is selected appropriately. The nodes are not attached with restricted battery power, which means they cannot be recharged or replaced.

Multi-Media WSN: The sensor nodes are arranged in a pre-planned fashion. It consists of number of low-cost sensor nodes attached with camera, microphone, etc. The interconnection with each other in the network over wireless connection for data retrieval, process analysis, correlation, and compression of data obtained from nodes. The main constrain in this type of WSN is high requirement of bandwidth and energy consumption, QoS requirement, processing data methods, compressing technique and cross-layer design and development. The network performance can be improved by processing, filtering, and compression significantly.

Mobile WSN: The sensor nodes are moving around their surrounding environment and communicate with each other. The nodes have functionality such as sensing, processing and communicating the data to base station with a high degree of coverage and connectivity. A mobile WSN has the significant characteristics that the mobile node is communicated to another mobile node when they are within the same network and gather information from each other. The Challenge includes network pattern, localization method, path selection organization technique, monitoring and control methodology, surrounding area, energy requirement, data processing and economic maintenance. The applications of M-WSN are environment monitoring, target monitoring and stolen recovery operation.

3.4.2 Mobile WSN Based Train Tracking Solution

WSNs are wireless network of distributed and autonomous devices. The sensor is used to cooperatively monitor the environment and infrastructure. One of the on-board WSNs for railway application is shown in Figure 3.4. The railway sensors which are used to monitor speed, location and direction of train are under the umbrella of Micro-Electro-Mechanical Systems (MEMS).

WSN needs to monitor the continuous real-time data. Each sensor network consists of sensor, microcontroller, data transmitter or receiver, antenna and power supply. The topology of WSN depends on real-time application. The research of the mobility (On-board) sensor node, where sensor node attached to locomotive or rail wagon is adopted. The data for movable network are combined with GPS coordinate and monitor the whole track line of train journey. The base station acts as gateway for data transmission to remote server. The sensor node transmit data to base station using small range communication such as Wi-Fi or Bluetooth and long-range communication such as GPRS, GSM or satellite to transmit collected data back to server at control center. The base station has a powerful processor and memory than the sensor node to transmit and receive data from multiple sensors. But the train monitoring system has a battery-powered sensor node in carriage and base station is located in train. Hence, the power is available for node and base station is obtained from train engine.

One main issue with movable sensor monitoring is communication, because of the mobility of sensor. The movable rail monitoring system transmits

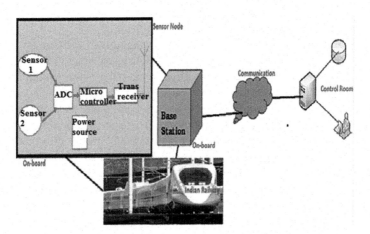

FIGURE 3.4   On-board WSN in train tracking application.

data from the movable node to the static node when the train stops in station. Then movable nodes will form a network with a base station within network. The nodes form network on train as base station is attached to the train, later transmits data to control Center through GSM-R. The stationary sensor node has less communication transmission capacity and area than mobile nodes.

Now by analyzing how data transmissions take place in tunnel when the base station and WSN are in movable mode. But the admitted fact is that it is difficult to collect and process data for transmission. GSM-R signals are available just outside the tunnel and they connect the gateway node to transmission computer directly with help of an Ethernet cable. Hence, no need for repeater unit, as it will reduce the chance of data loss, by connecting RF receiver unit to laptop for transmission of data to the server.

3.5 POSITIONING RAIL ACCURATE COMMUNICATION HIGHWAY IDENTIFICATION MODEL

The sensor-based, on-board train tracking model called **"Positioning Rail Accurate Communication Highway Identification"** (PRACHI) is viewed in Figure 3.5 which measures kinematics parameters and also presents situation values of moving train. The model measures location identification of train both in SVE and SLVE with meter level precise and accurate measurement. Other enhanced services include early warning signal on collision, track failure detection, speed restriction, warns derailment caused due to high speed.

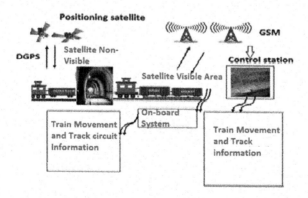

FIGURE 3.5   View of on-board sensor-based train tracking system.

This model supports train tracking and monitoring in both environments with the greatest accuracy to 95%–98% or better.

Higher requirements are set for safety and availability of train and track real-time information is achieved using our model. The basic work is obtaining real-time train identification and monitoring using DGPS-WSN-RFID based technology. For processing the data is transmitted to central control unit through the GSM network. Integrating two or more wireless technologies becomes a hot topic in the communication technology-based market.

Figure 3.6 depicts the block diagram of integrating system with three technologies. The wireless technologies have a wide variety of real-time applications in Internet of Things (IoT) vision. The integration of DGPS, WSN and RFID technologies is proposed for constant observing of train movement in both SVE and SLVE with greater tracking accuracy. The train movement is continuously tracked in satellite visible areas with help of DGPS technology and in low satellite visible areas with help of WSN and RFID integration technology. The real location data of moving train, speed, traffic condition and rail-track condition are helps train to avoid collision and derailment as well.

Designing novel approaches by integrating WSN and RFID systems for continuous tracking train movement in tunnel. In general, the WSN has only sensing surrounding condition than detecting a particular object. For identification and detection of moving train in tunnel can be carried out using RFID. Hence, it is necessary to combine RFID and WSN technology so that it gives a solution for moving train identification, detects and senses the condition of train equipped with sensors and enabled RFID tag. The two technologies are complementary to each other because they were originally designed with function such as identify, detect, sense and communicate.

The proposed system combines WSN and RFID wireless technology to compensate the limitations of each other. WSN provides good communication

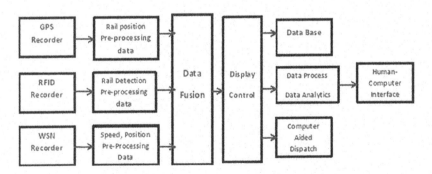

FIGURE 3.6 Block diagram of the wireless integration system.

coverage but with low tracking accuracy due to noise measurements. On the other hand, RFID technology provides very precise location information but limited coverage. The appropriate combination of technologies could be good plan in building positioning and tracking system with increased tracking accuracy and availability in satellite low visible environment (tunnel).

The data obtained from DGPS-RFID-WS module are fused and sent into database with help of a GSM network as a medium of communication. The proposed model is built in such a way that, possibility of signal failure due to one or two systems then continuous controlling and monitoring is done using a third system alone. The system also provides a web-based application interfaced with Google Map to display all transmitted information to end user. The central control system is responsible for processing and handling all the position information received from the train tracking model (**PRACHI**) through the GSM network. The server automatically updates the database with the latest position, speed, direction, train ID and name information of each train and also track condition. The accuracy of this information is very important to ensure the smooth functioning of railway service as well as to optimize resource planning.

Locomotive Tracking in Satellite Visible and Low Satellite Visible Area

4

4.1 INTRODUCTION

Real-world military and civil applications are in need of detection of moving target in remote location. In defense application, tracking is mostly from satellite-based navigation where the target's position is updated continuously. Another most obvious use is tracking of terrorist van for locking and vanishes easily. In real time, the object monitoring concept is well built-in fleet management, building security, etc. The moving object monitoring technology is used in a number of applications to make the system more automated and reduce the human intervention.

From past research and recent studies, it is concluded that the integrated system provides an acceptable accuracy with high dynamic information during GPS availability. However, system accuracy degrades in the tunnel with the absence of GPS, when the tracking solution is taken from a standalone system. To obtain the tracking information accurately in tunnel, it is better to integrate two or more systems. The satellite-based tracking real-time applications

DOI: 10.1201/9781003294016-4

for train monitoring exist everywhere. But these systems are facing many failure issues because of lack of communication between transmitter and receiver in black territory areas such as tunnels, underpass bridges, forests, etc. The communication failure mode in tunnel has created a promising challenge to the research community to prove a smart train monitoring system.

The chapter presents additional information on tracking surveillance algorithm and relevant software simulation used for implementation. We also briefly discussed the advantages of using the Discrete Kalman Filter in solving real-world problems and the WSN-RFID based train monitoring system. Finally, the chapter also presents a sensor model matching control system, tracking controller design, followed by quadratic optimal controller design based on Liapunov approach.

4.2 KINEMATICS UPDATE STATE HYPOTHESES INFORMATION SURVEILLANCE MODEL

Figure 4.1 shows the block diagram of the train surveillance model using DGPS navigation system. Differential GPS is a method of eliminating errors in GPS receiver to make the output more accurate. Unlike GPS, differential GPS is capable of providing high dynamic information with excellent tracking performance. Generally, DGPS is operated based on two receivers namely mobile and stationary receivers. Where the satellite position is calculated by mobile receiver and measured position signal to timing signal conversion is done by a stationary receiver in coordination with base station. Once the data are collected from satellites, pre-processing is done before the data transferred to Human Machine Interfacing (HMI) unit.

While designing for target tracking, the assumptions are listed below:

1. Constant state model with acceleration as state, often using position and rate measurement. In this model, acceleration is considered to be constant and the vehicle velocity remains constant for zero acceleration.
2. Constant state model with acceleration is considered as zero mean random input. The Kalman filter will cancel out zero mean acceleration as an input to state equation when acceleration is an input noise. In this method also, the vehicle velocity will tend to be constant.

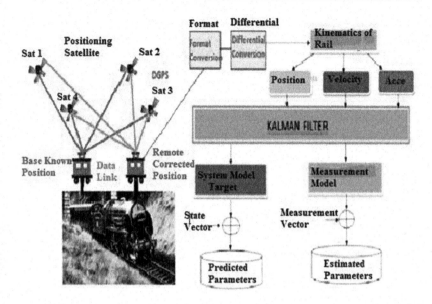

FIGURE 4.1   Train survelliance model using DGPS.

In any practical tracking model to track moving objects, then velocity as well as acceleration are considered as changing variables. Hence, it is preferable to estimate present and past acceleration. Similarly, by estimating the acceleration at each step with three past position measurements and this estimation methodology will carry the state transition matrix for time-varying model. As one would expect the estimates for train's velocity are clearly less accurate than estimate for train's position as the positions are observed directly and velocities indirectly. If train velocities were also observed not only velocity estimates would get more accurate, but also train position as well.

4.3 MANY TRACKING ALGORITHMS ONE FILTER- THE FUTURE WITH KALMAN FILTER

Many different algorithms have been proposed for target tracking, including two points extra polar, Wiener filter, G-H filter and Kalman filter. Each algorithm has strengths in certain environments and weaknesses in other.

Hence, the tradeoff between computation time and tracking accuracy is to be considered.

To design Discrete Kalman Filter (DKF) for robust train tracking in satellite visible areas is a major boost for any model. The filter can favor the algorithm that is most applicable to satellite visible environment by decreasing its measurement noise variance. We have selected DKF methodology because of its accurate position estimation. It takes care of missing and noisy measurements and continuously provides the best estimation depending on the available measurement value.

4.3.1 Why We Prefer Kalman Filter for Our Designing

The Kalman filter theory has been one of the best subjects for interesting research and real-time application, particularly in the hot area of target tracking. The smart solution in cloud-based technology for a digital filter is becoming more realistic because of its relative analysis, easy processing and optimizing nature of filter. Today it is assumed to be a promising algorithm used in the field of guidance, navigate, aircraft, missile and space vehicle tracking, etc. This chapter presents improvement in Kalman filtering applied to train tracking.

It is proposing to calculate the Gaussian distribution of a dynamic state of system. The recursive nature of this filter is one of the very appealing features and it makes its practical implementation much more realistic for the implementation of each designed data is directly proportional to each estimate. It also recursively conditions the current estimate on all past measurements. The filter is also called as Recursive Predictive Filter (RPF) because it is with state-space technique and recursive algorithm. If the system is disturbed by some noise then the filter is maintained estimation process with noise measurement as feedback.

1. The same filter is used for a variety of maneuvering and non-maneuvering target environments. Gain coefficients are computed dynamically.
2. The gain coefficient is computed dynamically for varying and missed detection.
3. It provides an accurate measure solution for covariance matrix, process model and gating model with better implementation.

Hence, the DKF algorithm is the foundation of this research because of its adaptability on digital computer implementation of prediction and estimation process using state-space model. The model equations are classified as (1) **State equation** which describes the moving object dynamics. (2) **Measurement equation** which determines state vector.

4.3.2 Designing Tree of DKF

The filter is acknowledged for Prediction Update (PU) and Measurement Update (MU) and Gain Matrix (GM) process by using feedback controller for noise measurement criteria. As such, equations for Kalman filter fall into two groups:

1. **PU:** At time $i+1$, the filter predicts the current state and variance depending on information at time i. This is also known as Time Update (TU). It is calculating the current state and error covariance estimated value to get the previous estimate value.
    1. **Current State Prediction (CSP):** $\hat{x}\left(i+\frac{1}{i}\right) = A\hat{x}\left(\frac{i}{i}\right)$
    2. **Prediction Covariance (PC):** $\tilde{p}\left(i+\frac{1}{i}\right) = A\tilde{p}\left(\frac{i}{i}\right)A^{T} + M(i)$
2. **MU:** At time $i+1$, the filter provides the current state and variance using future state and observed values $Y(i+1)$. The equations are very useful for measurement values which are in a feedback form. This equation is responsible for improved posteriori estimate.
    1. **Current State Estimate (CSE):** $\hat{x}\left(i+\frac{1}{i+1}\right) = \hat{x}\left(i+\frac{1}{i}\right) + K\left[y_i - C\hat{x}\left(i+\frac{1}{i}\right)\right]$
    2. **Estimation Covariance (EC):** $\hat{p}\left(i+\frac{1}{i+1}\right) = [I - KC]p\left(i+\frac{1}{i}\right)$
3. **GM:** It determines the relationship between Mean-Squared Estimation Error (MSEE) and Kalman gain.

Kalman gain (KG): $\quad K = \hat{p}\left(i+\frac{1}{i}\right)C^{T}\left[C\tilde{p}\left(i+\frac{1}{i}\right)C^{T} + N\right]^{-1}$

The TU equation considers a predictor equation, while the MU equation as corrector equation.

4.3.3 Theoretical Testing of Kalman Filter

Kalman filter is also tested theoretically by the following process. The steps are as follows;

1. State Process Model; the following assumptions are to be made.
   - State variable reduced to scalar
   - Assuming constant model (A=1)
   - Control variable B, U=0
2. Measurement Process Model; Following assumption is made.
   - Measurement matrix C=1
3. Noise Modeling; Following assumptions are to be made.
   - Assume Gaussian white noise
   - Process is scalar (N=n, P=p, M=m)
4. Test the filter; Following state equations are to be tested.
   - State Prediction: $\hat{x}\left(i+\frac{1}{i}\right) = A\hat{x}\left(\frac{i}{i}\right)$
   - Prediction Covariance: $\tilde{P}\left(i+\frac{1}{i}\right) = A\tilde{P}\left(\frac{i}{i}\right)A^T + M(i)$
   - State Estimate: $\hat{x}\left(i+\frac{1}{i+1}\right) = \tilde{x}\left(i+\frac{1}{i}\right) + K\left[y_i - C\hat{x}\left(i+\frac{1}{i}\right)\right]$
   - Estimation Covariance: $\hat{p}\left(i+\frac{1}{i+1}\right) = [I - KC]\hat{p}\left(i+\frac{1}{i}\right)$
   - Kalman gain: $\hat{p}\left(i+\frac{1}{i}\right) = C^T\left[C\tilde{p}\left(i+\frac{1}{i}\right)C^T + N\right]^{-1}$
5. Initialize the model; the following initial parameters are to be assumed.
   - System Noise M=0.0001
   - Measurement Noise N=0.1
   - X0=0 and P0=1000

Table 4.1 shows the summaries of theoretical value of Kalman filter designing.

4.4 MODELING ASSUMPTION FOR DGPS MEASUREMENTS USING KALMAN FILTER

Designing model is focused on the varying acceleration parameters because of their accurate measurement. The constant acceleration model will make a change in velocity constant with sample time period T seconds. The part of problem formulation is concentrated mainly on subject when a train takes its journey in SLVE. Consider the train is moving in straight trajectory with constant velocity. Let x (i) and x' (I) are train position and velocity respectively. Let v (i) be measurement noise which observe the position of train. When the train is moving at constant speed $X''(i)=0$. The kinematic parameters such as

TABLE 4.1 Summary of real data

SL. NO		REAL DATA
1	Sensor Location	$S_1(0,0,0)$, $S_2(50,0,0)$
2	Standard deviation of positional measurement error (in mt.)	Std-x = Std-y = 0.0001
3	Field of View (in mt.)	$X_{min} = 0$, $X_{max} = 50$ $Y_{min} = 0$, $Y_{max} = 50$
4	Sampling Interval	1 second
5	Initial State Estimate Covariance	Positional Variance = 0.1(X, Y) Velocity Covariance = 0.0001(V_x, V_y)
	Initial Track Score	0.9
6	Process Noise Covariance	0.9* Measurement error covariance (R)
7	Pruning Threshold	0.01
8	Combining Similar Track	Combining similar track $N_D m^2$
9	Distance Threshold	10m

position, velocity and acceleration are considered and the state vector is represented as $X = [Position, Velocity, Acc]^T$ x, y, z.

Designing a state vector is of great interest with state of dynamic system, degrees of freedom and state vector variable. When the train is moving in SVE it offers 2-degree of freedom, distance and velocity. The model vector has predicted and posterior value of update values. The implementation of the Kalman filter requires prior knowledge of both process and measurement model. Let's now consider tracking the train moving in two-dimensional space with a sensor, which gives measurements of train's position in Cartesian coordinates x and y. In addition to position, the train also has state variables for its velocities and accelerations toward coordinate axes, x, y, x^1 and y^1.

In train trajectory estimation, it is necessary to identify behavior of model interest. The behaviors of model interest include turning, accelerating and braking. By suppressing the braking and turning, our research is choosing the model to represent a set of these behaviors and adding sophistication to the previous one. The model will estimate:

- Straight-line travel at constant velocity
- Travel with constant straight-line acceleration
- Travel with constant turning acceleration and constant speed. Here the train is modeled as traveling at an approximately constant velocity in magnitude and direction.

4.4.1 Kinematic Tracking State Model

Let us consider that the train is moving in a straight line with constant velocity. Designing the state model based on kinematic parameters of train is called tracking state model. The designing is carried out with a period from n to n+1 along X and Y coordinates. The position measurement along two coordinates are X P (n) and YP (n) and locomotive velocity measurement values in two directions are XV (n)and YV (n) respectively.

The train movement is described by the dynamic equations

$$x[i+1] = A x[i] + Bu[i] \tag{4.1}$$

Where x [i] is state vector, u [i] is input vector. x [i] will be defined as

$$x[i] = \begin{bmatrix} P(i) \\ V(i) \\ Acc(i) \end{bmatrix} \tag{4.2}$$

where P (i) is the position of the moving train, V (i) is its velocity and Acc (i) be acceleration. Consider step as input for the system. Let 'A' be transition matrix of system parameter and the GM of process noise measurement be 'B' with sampling duration of δt.

$$A = \begin{bmatrix} 1 & \delta t & \delta t^2/2 & \delta t^3/3 \\ 0 & 1 & \delta t & \delta t^2/2 \\ 0 & 0 & 1 & \delta t \\ 0 & 0 & 0 & 1 \end{bmatrix} \quad B = \begin{bmatrix} \delta t^3/3 \\ \delta t^2/2 \\ \delta t \\ 1 \end{bmatrix} \tag{4.3}$$

$$x[i+1] = A x[i] + Bu[i]$$

$$\begin{bmatrix} x(i+1) \\ x'(i+1) \\ y(i+1) \\ y'(i+1) \end{bmatrix} = A \begin{bmatrix} x(i) \\ x'(i) \\ y(i) \\ y'(i) \end{bmatrix} + B \begin{bmatrix} u_x(i) \\ u'_x(i) \\ u_y(i) \\ u_{y'}(i) \end{bmatrix} \tag{4.4}$$

By considering the tracking state model parameters for the duration is from i to i−1, while the transition and GM of system and noise measurement are not going change anymore.

$$x[i] = Ax[i-1] + Bu[i-1] \tag{4.5}$$

The moving object measurement noise covariance are representing in matrix form be 'M', Variance $\{u(i)\} = M$, and consider $E\{u(i)\} = 0$

4.4.2 Parameter Measurement Coordinate Model

The measurement model is common for all Kalman filters. The filters receive the same input, target position, and adopt the same coordinate frame expression of position in their state vector. Further, before it enters Kalman filter algorithm we carry out coordinate transformation from global 3-D form to 2-D form.

$$Y(i) = CX(i) + V(i) \tag{4.6}$$

where C is sensor output

$$C = \begin{bmatrix} 1 & \delta t & \delta t^2/2 & \delta t^3/3 \\ 0 & 1 & \delta t & \delta t^2/2 \\ 0 & 0 & 1 & \delta t \\ 0 & 0 & 0 & 1 \end{bmatrix} \quad \text{If } \delta t = 0 \quad C = \begin{bmatrix} 1 & 0 & 0 & 0 \\ 0 & 1 & 0 & 0 \\ 0 & 0 & 1 & 0 \\ 0 & 0 & 0 & 1 \end{bmatrix}$$

$$\begin{bmatrix} Y_y(i) \\ Y_x'(i) \\ Y_y(i) \\ Y_y'(i) \end{bmatrix} = \begin{bmatrix} 1 & 0 & 0 & 0 \\ 0 & 1 & 0 & 0 \\ 0 & 0 & 1 & 0 \\ 0 & 0 & 0 & 1 \end{bmatrix} \begin{bmatrix} x(i) \\ x'(i) \\ y(i) \\ y'(i) \end{bmatrix} + \begin{bmatrix} v_x(i) \\ v_x(i) \\ v_y(i) \\ v_y(i) \end{bmatrix} \tag{4.7}$$

For the measurement of noise that occurred in feedback are represent in matrix form be 'N' and consider $\{v(i)\} = N$ and $E\{v(i)\} = 0$.

4.5 ADOPTED ALGORITHM FOR KALMAN FILTER MODEL TESTING

The main assumptions are considered while drafting the algorithm. Equations (4.1) and (4.5) are tested theoretically by assuming the values of

- Constant Model with $A = 1$
- Control Variables are set to B, $U = 0$
- Measurement matrix $C = 1$
- Process is Scalar with $N = n$, $P = p$ and $M = m$.
- Assume Gaussian White noise.

With these values, the following algorithm (1) and algorithm (2) are tested to verify the modeling accuracy.

Adopted Algorithm (1): Prediction and Estimation Formulator

- Define train's kinematic updated values: $\hat{x}\left(i + \frac{1}{i}\right) = A\hat{x}\left(\frac{i}{i}\right)$, $A =$ transition matrix parameter
- Note the noise in feedback controller: $\tilde{p}\left(i + \frac{1}{i}\right) = A\tilde{p}\left(\frac{i}{i}\right)A^{T} + M(i)$, Assume System Noise $M = 0.0001$, $X_0 = 0$ and $P_0 = 1000$
- Consider train's State Estimation value: $\hat{x}\left(i + \frac{1}{i+1}\right) = \tilde{x}\left(i + \frac{1}{i}\right) + K\left[y_i - C\hat{x}\left(i + \frac{1}{i}\right)\right]$
- Consider Estimation Covariance: $\hat{p}\left(i + \frac{1}{i+1}\right) = [I - KC]\hat{p}\left(i + \frac{1}{i}\right)$
- Define Kalman gain: $K = \hat{p}\left(i + \frac{1}{i}\right)C^{T}\left[C\tilde{p}\left(i + \frac{1}{i}\right)C^{T} + N\right]^{-1}$, Measurement Noise $N = 0.1$

Algorithm (1) illustrates the Kalman filter prediction and estimation formulation. Similarly, algorithm (2) represents how the dynamics of train is calculated when it is traveling in a purely satellite visible area.

Adopted Algorithm (2): Position, Velocity of Train Updates

Kinematic parameters Value [$x_{prediction}$, $p_{prediction}$] = Predict(x, p, A, M)
$X_{prediction} = A*x$;

$P_{\text{prediction}} = A*p*A^1 + M;$

Estimated Value [Sub, T] = Dynamic $(x_{\text{prediction}}, p_{\text{prediction}}, y, C, N)$

$Sub = y - C*x_{\text{prediction}};$

$T = N + C*p_{\text{prediction}} *C^1$

Resultant Value $[x_{\text{KUSHI}}, p_{\text{KUSHI}}]$ = Dynamic @ update $(x_{\text{prediction}}, p_{\text{prediction}}, Sub, T, C)$

$K = p_{\text{prediction}} *H^1*T^1$

$X_{\text{KUSHI}} = x_{\text{prediction}} + K*Sub$

$P_{\text{KUSHI}} = p_{\text{prediction}} - K*T*K^1$

Nominal Parameters Used in Algorithm

- Location of Sensor - SL_1 – [0,0,0] SL_2 – [60,0,0]
- Error of Positional Measurement - x-direction = 0.0001 m
- Sampling duration – 1 second, First monitoring reading – 0.8
- Noise measurement of Process Covariance = 0.8 * Measurement error covariance
- First State Estimated Covariance values - Position Variance = 0.1 [x, y]
  Velocity Covariance = 0.0001 [x, y]

4.6 TEST ROUTE CASE STUDY

The case study of single-train test journey is considered. The test journey path simulation is based on rail line between Madgaon (Goa) to Honavar (Uttara Kannada) with an intermediate stop. Figure 4.2 depicts 12619 Mathysaganda Express realistic train trajectories operated between Madgaon to Honnavar. The graph shows six intermediate main station stops, three bridges and two tunnels along the route. The simulation input data for the case study are:

- Rail route length = 150 km
- Destination station = 150 km
- Intermediate stations = 6
- Train stop time = 3 minutes

The experiment is conducted to verify the tracking accuracy of a model. A railway tracking system provides train route along with kinematics information while ensuring train safety and monitoring train operation. The hard core of control objectives are train speed, train position, railway points and signaling

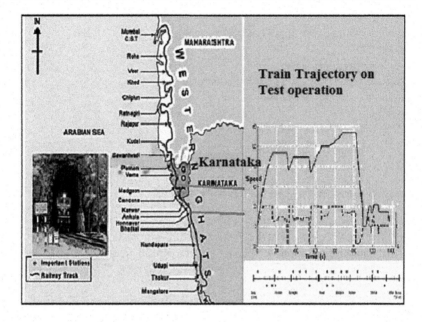

FIGURE 4.2   The scenario of single-train operating on test route.

aspects. When the train is in journey mode, DGPS measure traveled path and its related data to train on-board sensor system, which then determines the location and speed profile.

Today the fast evolution in automation-based technology is becoming a hot area as it supplies economical and feasible methodology for assessment. The single-train simulator is designed in MATLAB® and is feasible to analyze movement data on single railway lines with fixed stations. Train journey trajectory, tracking accuracy measurement output are plot when the simulation cycle is complete. Considering the train is moving in a straight way with constant velocity. Simulation is done using MATLAB® algorithm and Visual Kalman Filter Window. It provides a visual method to estimate the state of the process and removes noise from data. Following are some of the sample cases for true and estimated parameters using the Kalman filter design methodology.

4.6.1 Case 1: Position Identification of Moving Train Using Kalman Filter

As seen in Figure 4.3, it is an appeal that the simulation values are approximated to the real data. The simulation result shows that the train journey

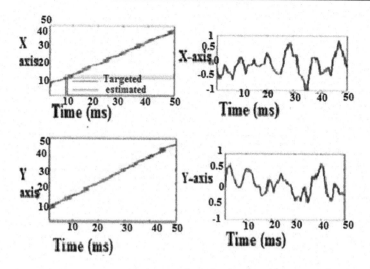

FIGURE 4.3   Position tracking along X-Y direction and related data loss.

takes 10 seconds less than the pre-defined traveling duration. This is very difficult to maintain modeling the train trajectory precisely in a real case. The train continued to travel until the actual speed reached the line speed value. However, the pre-defined strategy of the locomotive traveling between station stops, bridges as well as tunnel are monitoring continuously during this period. Hence, the average speed of simulated train journey data is slightly greater than the real data.

The initial traveling path is considered for run of 60 intervals in straight single line. The research design simulated model values are based on true journey conditions, and the estimated data are very close to true journey value. The X-coordinate positional state data loss and Y-coordinate positional state data loss are complementary to one another. With X-direction, the position state error (PSE) are estimated to be +0.5 at t=30 and −1 at t=35. Similarly, with Y-direction, the PSE is estimated to be −1 at 30 and +0.5 at t=35 respectively.

4.6.2 Case 2: Velocity Measurement of Moving Locomotive Using Kalman Filter

Figure 4.4 shows the velocity measurement of locomotives. From the graph, we observe that the simulated velocity values are approximated to the true velocity value and that the measurement noise is at least a considerable value. The velocity data shows the variation when the train is taking the journey

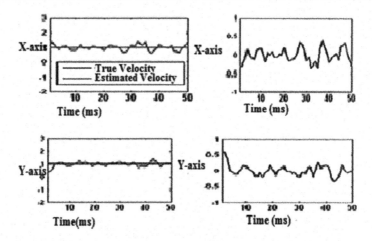

FIGURE 4.4 Velocity measurement along X-Y direction and data loss.

TABLE 4.2 Velocity-position tracking error estimation

CASE	VELOCITY	POSITION
	$V_x V_y$	$P_x P_y$
U(k)=0.7913 J=0.5	0.2–0.2	0.4–0.4
U(k)=−0.2087 J=1.8	0.5–0.6	0.6–0.5

in X-direction than when the train is moving in Y-direction as expected. It is concluded that the simulated velocity data is approximately nearer to the pre-defined velocity value of the model. Table 4.2 describes the real datasheet.

4.7 WSN-RFID BASED TUNNEL SURVEILLANCE INTEGRATION MODEL

The function of the surveillance model is to identify the allocated area, detect the moving object and next communicate the data to the central control unit. Hence, a surveillance system is in need for continuously observing the train journey in a tunnel. To overcome the associated limitation with space-based radio navigation technologies, it is more advantageous to integrate two or more wireless sensors system to increase tracking accuracy are preferred.

The research highlights two or more fusion technology because system failure will compensate for each other. The information obtained from wireless sensors is helping in improving the train and its traveling path detection in a tunnel. Thus, in a continued effort to improve the tracking accuracy, we design Wireless Sensor (WS) and Radio Frequency Identification based integrating model. Figure 4.5 depicts the WSN-RFID based tunnel surveillance heuristic integration model.

The novel approaches to integrating WSN and RFID systems for continuous tracking train movement in the tunnel is explained here. Generally, WSN has only sensing surrounding conditions than detecting a particular object. For identification and detection of moving train in a tunnel can be carried out using RFID. Hence it is necessary to combine RFID and WSN technology so that it gives a solution for moving train identification, detects and senses the condition of train equipped with sensors and enabled RFID tag. The two technologies are complementary to each other because they were originally designed with a function such as identify, detect, sense and communicate.

The proposed WSN-RFID based positioning system is used to compensate for the limitations of each wireless technology. WSN provides good radio

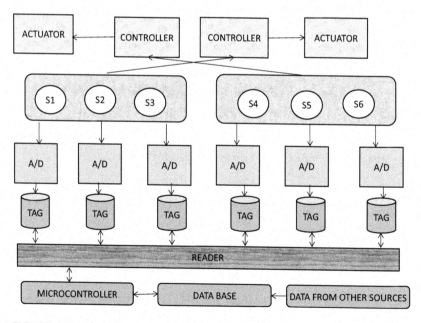

FIGURE 4.5 Block diagram of sensor integration model.

coverage but with low tracking accuracy due to high noise measurements. On the other hand, RFID technology provides very precise positioning information but limited coverage. The appropriate combination of technologies has good tracking accuracy and availability in building the SLVE model. The inter-communication within the sensors network is analyzed using distributed control algorithm. The base station acts as a controller which can supply the information to sensor which is arranged in each subsystem. Similarly, the RFID tag is equipped with a wireless module for communicating all train journey movement data. Later data are transmitted to the reader which is then transmitted to the on-board system located in the train.

4.8 SENSOR MATCHING CONTROL ALGORITHM FOR DI-SENSOR MODEL

The methodology to find sensor transfer function using Di-Sensor $\alpha \beta$ model-based SMC (Sensor Matching Control) algorithm for better tracking accuracy is explained in a flow diagram. The Di-Sensor $\alpha \beta$ model is designed with the integration of the α model based on DGPS technology and the β model based on WSN technology. In real-time application, tracking the train in a tunnel leads to variation in sensor tracking accuracy performance. This model can track the locomotive trajectory easily as well as avoid degradation in tracking accuracy performance.

Let us consider α- model (RFID) and β model (WSN) with A(z) and B (z) are i^{th} and j^{th} degree polynomial for sensor stable transfer function respectively. The design modeling of sensors with systems is analyzed by a polynomial equation for the desired stable transfer function. The transfer function of the two-sensor model for a closed-loop system is calculated as,

$$Y(z)/U(z) = \text{WSN-based } \beta \text{ polynomial B}(z)/\text{RFID-based } \alpha \text{ polynomial A}(z)$$
$$(4.8)$$

With the foundation of matching the two-control system, this research proposes polynomial equation-based transfer function calculation.

Let H_1 (z) be the system transfer system, the design process continues as,

$$H_1(z) = H(z)/A(z) \; \alpha\text{-}\mathbf{RFID \; Model} \text{ and } H_2(z) = H(z)/B(z)\beta\text{-}\mathbf{WSN \; Model}$$

Thus $Y(z) / R(z) = K_{Model} = B_J(z) / A_I(z)$ $\hspace{2cm}$ (4.9)

Let us consider, F(z) be the controllable model with $(n-1)^{th}$ degree polynomial.

$$P(z)A(z)+Q(z)B(z) = F(z)B(z)H_1(z) \qquad (4.10)$$

The control system analysis for sensor is represented as,

$$U(z) = -\left[P(z)/F(z)*U(z)-U(z)+Q(z)/F(z)*Y(z)\right]+V(z) \qquad (4.11)$$

$$V(z) = P(z)/F(z)* U(z)+Q(z)/F(z)*Y(z)$$

$$= P(z)/F(z)* A(z)/B(z)* Y(z)+Q(z)/F(z)* Y(z)$$

$$\left[U(z)=A(z)/B(z)*Y(z)\right]$$

$$Y(z)/V(z) = F(z)B(z)/P(z)A(z)+Q(z)B(z)$$

$$= F(z)B(z)/F(z)B(z)H_1(z) = 1/H_1(z)$$

As $V(z) = K_{Model}H_1(z)R(z),$

$$Y(z)/R(z) = Y(z)V(z)/V(z)R(z) = K_{Model} H_1(z)/H_1(z)$$

Hence

$$V(z)/R(z) = K_{Model}H_1(z) \qquad (4.12)$$

The characteristic of equation (4.7) shows the improvement in stability of the system model. Now, the sensor matching model is ready for tracking the train in tunnel.

4.9 METHODOLOGY TO ANALYSIS SENSOR MATCHING CONTROL DESIGN FOR TRAIN TRACKING

When the train is moving in tunnel, the navigation information is reduced to a certain level and next it is difficult to measure velocity and speed profile. To overcome this limitation, the research proposed an integration model of RFID-WSN to detect the train and measure its kinematic parameters. The proposed model is analyzed using a digital control system by building a type 1

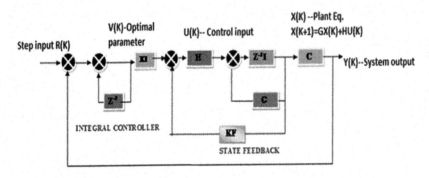

FIGURE 4.6   Tracking controller model.

Servo System. It is assumed that the designed RFID-WSN model is equipped on the train (strictly not on track) and controls the movement of the train. Figure 4.6 shows the controller model where the plant consists of sensor state feedback and an integrator in closed-loop fashion and the control force is applied to the moving train.

Consider the journey in step fashion with step input of R (k) and displacement Y (k). The design parameters are (1) Integral controller Gain constant K_I (2) State feedback controller Feedback GM KF (3) State of system X (k) (4) Control input U (k) (5) System state matrix G [n*n matrix] and (6) Control matrix H [n*r matrix].

4.10 TRAIN TRACKING PROBLEM FORMULATION

Quadratic Optimal Control (QOC)-based Wireless Tracking Controller (WTC) is designed with Liapunov approach. The Liapunov function has a direct relation with Performance Index (PI). With variation in PI value to control inputs will calculate tracking data loss. Let U(k) be control input (step input) is applied to the WTC. Let us now consider QOC design, where the process is bound with $M = \infty$

The system design equations are modified according to our requirement.

$$x(k-1) = Ax(k) + Bu(k) \tag{4.13}$$

$$y(k) = Cx(k) \tag{4.14}$$

$$v(K) = v(k+1) + r(k) - y(k) \tag{4.15}$$

$$v(k-1) = v(k) + r(k-1) - y(k-1)$$
$$= v(k) + r(k-1) - CA\, x(k) - CB\, u(k) \tag{4.16}$$

The control inputs or control forces are applied to the train,

$$u(k) = K_I v(k) - K x(k) \tag{4.17}$$

where
K_I = Gain constant and K = Feedback GM

Let us consider the sensor state model for the kinematic parameter of locomotive for two coordinates (X, Y) with duration $K - 1$ to K. Based on this assumption the model design is

$$\begin{bmatrix} X(K-1) \\ V(K-1) \end{bmatrix} = \begin{bmatrix} A & 0 \\ -CA & 1 \end{bmatrix} \begin{bmatrix} X(K) \\ V(K) \end{bmatrix}$$

$$+ \begin{bmatrix} B \\ -CB \end{bmatrix} U(K) + \begin{bmatrix} 0 \\ 1 \end{bmatrix} R(K-1) \tag{4.18}$$

Where the system parameters are,

$$A = \begin{pmatrix} A & 0 \\ -CA & 1 \end{pmatrix} \qquad B^\wedge = \begin{bmatrix} B \\ -CB \end{bmatrix} \qquad K^\wedge = [K, -K_r]$$

Now consider QOC where the process is continuous with $K = \infty$.

$$\begin{bmatrix} X(\infty) \\ V(\infty) \end{bmatrix} = \begin{pmatrix} A & 0 \\ CA & 1 \end{pmatrix} \begin{bmatrix} X(\infty) \\ V(\infty) \end{bmatrix} + \begin{bmatrix} B \\ CB \end{bmatrix} U(\infty) + \begin{bmatrix} 0 \\ 1 \end{bmatrix} R(\infty) \tag{4.19}$$

From equation (4.8), the tracking error accuracy is

$$\begin{bmatrix} X_e(K-1) \\ V_e(K-1) \end{bmatrix} = \begin{bmatrix} A & 0 \\ -CA & 1 \end{bmatrix} \begin{bmatrix} X_e(K) \\ V_e(K) \end{bmatrix}$$

$$+ \begin{bmatrix} B \\ -CB \end{bmatrix} U(K) + \begin{bmatrix} 0 \\ 1 \end{bmatrix} R(K+1) \tag{4.20}$$

The step input R (K),

$$R(K) = R(K-1) = R$$

By defining, $X_e(K) = X(K) - X(\infty)$ and $V_e(K) = V(K) - V(\infty)$
Then error equation (4.8) becomes

$$\begin{bmatrix} X_e(K+1) \\ V_e(K+1) \end{bmatrix} = \begin{bmatrix} A & 0 \\ -CA & 1 \end{bmatrix} \begin{bmatrix} X_e(K) \\ V_e(K) \end{bmatrix} + \begin{bmatrix} B \\ -CB \end{bmatrix} U_e(K) \qquad (4.21)$$

$$U_e(K) = -K X_e(K) + K_I V_e(K) = -[K - K_I] \begin{bmatrix} X_e(K-1) \\ V_e(K-1) \end{bmatrix}. \qquad (4.22)$$

The significance of the above equation is that the control input mainly depends on state gain as well as feedback factor which in turn degrade tracking inaccuracies.

4.11 DECISION LOGIC FOR PI AND TRACKING ACCURACY

The suitable PI is chosen for designing an optimal control system. Based on the system requirements, the criteria of PI should meet the condition to calculate the stability of a robust system. Hence, for system optimization, the control vector U (K) depends on any variation in the PI values.

According to optimal control law,

$$U(K) = -KX(K) \qquad (4.23)$$

The PI related with steady-state optimal control law

$$J = 1/2 \sum_{k=0}^{\infty} [X*(K)EX(K) + U*(K)FU(K)] \qquad (4.24)$$

where
   $E = n \times n$ Positive semi-definite Hermitian matrix.
   $F = n \times r$ Positive semi-definite Hermitian matrix

In an optimization problem, the value of E and F matrices has an effect on PI, state vector, control vector and final state.

4.12 QUADRATIC PERFORMANCE CONTROL PROBLEM ANALYSIS USING LIAPUNOV METHOD

A number of optimized methods are suggested for solving control problems based on the relationship between PI and Liapunov functions. Consider an optimal control problem where plant equation (4.8), optimal control law (4.18) and PI (4.19) are optimized as,

$$X(K-1) = AX(K) + B[-KX(K)], \text{ Then } X(K-1) = X(K)[A-KB] \quad (4.25)$$

The relationship between PI and control law exists as,

$$J = 1/2 \sum_{k=0}^{\infty} \left[X^*(K)EX(K) + U^*(K)FU(K) \right]$$

$$= 1/2 \sum_{k=0}^{\infty} [X^*(K)\{E + K^*FK\}]X(K)] \quad (4.26)$$

For parameter optimization, matrix (A-KB) is stable and Eigen value is within the unit circle. For equation (4.8), Liapunov function is defined as,

$$V\{X(K)\} = X^*(K)\ PX(K)$$

where P is positive definite Hermitian matrix and whose derivative is negative definite.

$$dV/dk(X(K)) = V\{X(K-1)\} - V\{X(K)\}$$

$$= X^*(K-1)\ PX(K-1) - X^*(K)\ PX(K) \quad (4.27)$$

Simplifying the below Equations,

$$X^*(K)\{E + K^*FK\}X(K) = -[X^*(K-1)PX(K-1) - X^*(K)PX(K)] \quad (4.28)$$

$$X^*(K)\{E + K^*FK\}X(K) = X^*(K)[(A-BK)^* P[(A-BK)-P]X(K)] \quad (4.29)$$

By comparing,

$$E + K^*FK = -[(A-BK)^* P[(A-BK)-P]] \quad (4.30)$$

$$E + K*FK + A*PA - A*PBK - B*K*PA + B*K*PBK - P = 0$$

$$E + A*PA - P + \left[(F + B*PB)^{1/2} K - (F + B*PB)^{-1/2} B*PA \right]$$

$$\left[(F + B*PB)^{1/2} K - (F + B*PB)^{-1/2} B*PA \right]$$

$$- A*PB(F + B*PB)^{-1} B*PA = 0 \tag{4.31}$$

PI and the value of matrix K are minimized,

$$\left[(F + B*PB)^{1/2} K - (F + B*PB)^{-1/2} B*PA \right]\left[(F + B*PB)^{1/2} K - (F + B*PB)^{-1/2} B*PA \right]$$

Is zero or is non-negative.

$$(F + B*PB)^{1/2} K = (F + B*PB)^{-1/2} B*PA,$$

then $K = (F + B*PB)^{-1} B*PA$ \hfill (4.32)

$$P = E + A*PA - A*PB(F + B*PB)^{-1} B*PA \tag{4.33}$$

By using identity matrix, I, matrix P is calculated as,

$$P = E + A*P\left(I + BF^{-1}B*P\right)^{-1} A$$

The minimum and maximum value of PI is obtained,

$$J = 1/2 \sum_{k=0}^{\infty} \left[\left[X*(K)\{E + K*FK\}\right] X(K) \right]$$

$$J = 1/2 \sum_{k=0}^{\infty} \left[X*(K+1)PX(K+1) - X*(K)PX(K) \right] \tag{4.34}$$

From the equation, it is concluded that for better optimization PI and control input are equally important.

Adopted Algorithm: Working of the PI Logic

- Insert an integrator in feed-forward path.
- The control input 'u (k)' is applied to the moving train.
- Position signal outcome 'Y (k)' is connected back to input.
- Plant equation of discrete time control system

$$X(k+1) = G X(k) + H U(k)$$

- Select the optimal control vector with respect to PI

$$J = \tfrac{1}{2}X^*(N)SX(N) + \tfrac{1}{2}\left[X^*(k)QX(k) + U^*(k)RU(k)\right]$$

- Find the feedback GM K
- Apply optimal control vector U (k) in step
- Find the integral gain constant K_I
- Display the output of integrator.

With variation in PI value to control inputs will calculate tracking data loss. For the operation of the wireless controller in all the cases, control input (step) and determine how much data loss and tracking accuracy recovered in each process. For checking the performance of the WSN-RFID based tracking controller model, experiments are conducted for different values of PI and control inputs.

4.12.1 Case 1: For Optimal Control Input U (K) = −0.7913, Performance Index J = 0.5

The tracking accuracy calculation with variation in control input graph in Figure 4.7 highlights that state velocity readings are almost nearer to the pre-defined value when train trajectory is considered along X and Y coordinates respectively. Figure 4.8 highlight that position readings are almost nearer to the pre-defined value when train trajectory is considered along X and Y coordinates. By increasing the value of PI to a certain value will make improvement in response. It also shows the production of 0.006% inaccuracy in data loss.

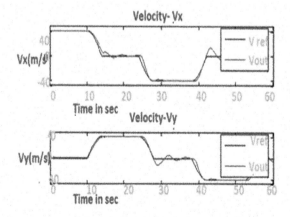

FIGURE 4.7   Velocity profile and its tracking error of train.

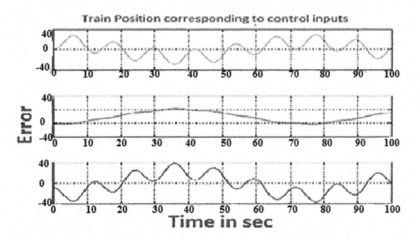

FIGURE 4.8   Position profile and its tracking error of moving train.

The position destination value is not approximated to the reference value but the journey is in opposite direction.

4.12.2 Case 2: For Optimal Control Input U (K) = −0.2087, Performance Index J = 1.8

The tracking accuracy calculation of velocity and position value with variation in control input highlights that the estimated state velocity readings are almost nearer to the pre-defined value when train trajectory is considered along X and Y coordinates respectively. The position readings are almost nearer to the pre-defined value when train trajectory is considered along X and Y coordinates. Increasing the value of PI to a certain value will make improvement in response. It also shows the production of 0.004% inaccuracy in data loss. The position destination value is not approximated to the reference value but the journey is in the opposite direction

4.12.3 Train Tracking with Velocity and Position Error Estimation

The comparison result shows that tracking data loss occurred in both measurements, which seems to be the same value, but path traveled values in the opposite direction. It is also summarizing that tracking performance data loss is more in type 2 than in class 1.

Train Trajectory Optimization Based on Di-Filter Theory

5

5.1 INTRODUCTION

The surveillance technique plays an important role in moving object tracking in major sectors. But till today it is facing a lot of technical awareness in safety control system. An end-to-end smart solution is needed to strengthen the railway sector in terms of safety measurements. Generally, the function of a surveillance model is believed to identify the allocated area, detect the moving object and next communicate the data to the central control unit. Hence, a surveillance system is in need for continuously observing the train journey in a tunnel. To overcome the associated limitation with space-based radio navigation technologies, it is more advantageous to integrate two or more wireless sensor systems to increase tracking accuracy.

DOI: 10.1201/9781003294016-5

5.2 OPTIMAL DIFFERENTIAL CORRECTION SOLUTION FOR STANDALONE GPS

The standalone GPS is not able to monitor detection of object in signal shading area such as in tunnel, thick forest area, urban canyon, mountain, valley etc. Suppose a vehicle is moving along a closed bridge, then GPS signals are not available for some time. This is because it suffers from line of sight and multipath loss, which results in lesser tracking accuracy. No position will be obtained in a tunnel. Hence, a smart solution is needed where the journey of train during the period of GPS coverage is re-constructed. The estimation error has correlation for real time application at restricted area. For improving the train tracking accuracy instead of using GPS, it is better to have differential GPS in order to reduce the measurement errors.

The part of the chapter proposed the improvement in train identification and monitoring system in two different areas, such as (1) SVE (2) SLVE. Hence, with the IOT-based intelligent solution, an accurate and effective continuous tracking capability is essential for rail sector. The following are our work toward building train tracking in daily need application.

The sensor-based on-board train trajectory monitoring model **"Sensor Accuracy Remote Access Surveillance Wireless Automatic Tracking Heuristic Innovation"** is designed. The model was actually developed with DGPS to navigate moving train journey in two different environments such as SVE and SLVE. The model measures kinematics parameters like position, velocity and other surrounding conditions values of the train. Figure 5.1 describes the block diagram of train trajectory measuring model.

Differential GPS is a method of eliminating errors in GPS receiver to make the output more accurate. Unlike GPS, Differential GPS is capable of providing high dynamic information with excellent tracking performance. Generally, DGPS is operated based on two receiver namely mobile and stationary receivers. Where the satellite position is calculated by mobile receiver and measured position signal to timing signal conversion is done by stationary receiver in coordination with base station. Once the data are collected from satellites, pre-processing is done before the data transferred to Human Machine Interfacing (HMI) unit. The DGPS tracking module detects the train location and communicates the data to main control unit. The on-board system data will help the train operator to take decisions. The position and speed profile of the train help the main system to manage with precautionary measure of safety. In the 21st century, DGPS takes on greater significance in improving tracking accuracy advancement.

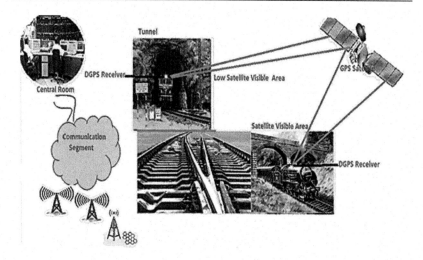

FIGURE 5.1   Real time scenario of train trajectory tracking.

5.3 INTERACTING MULTIPLE MODEL (IMM) ALGORITHM FOR DI-FILTER MODEL

The two-tuned filter theory explains how Kalman filter uses the arrangement of filters in order to obtain resultant tracking values. The research proposed a Di-filter model for SVE and SLVE, respectively. In real-time tracking application, the Di-filter model is used to calculate the train journey from SVE to SLVE two different areas. The two operating modes where the control characteristics of Di-filter like transmission channel, gain and measurement noise are selected in SVE and SLVE respectively. To analyze this parameter, the well-known IMM algorithm is considered. For the analysis purpose, the research proposes that "M_1" for SVE and "M_2" for SLVE are operating modes of filter. The IMM algorithm-based mixed model is shown in Figure 5.2.

IMM algorithm is modular design with estimator that is in recursive nature. It consists of four main processing rules, such as initial interaction, different mode-conditioned and processing filtering, testing probability evaluation, combined state transition and covariance estimation. The mixed mode system works on filter (mode-1) and predictor (mode-2) stages. When the train is moving in SVE, the system acts in mode-1 and automatically changes to mode-2 when the train is in a low satellite visible area. The experiment was developed for changing mode also. This experiment shows how the mixed

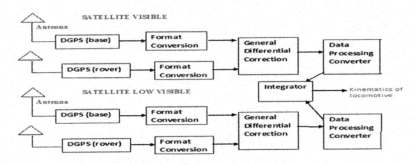

FIGURE 5.2   Block diagram of train trajectory innovation model.

system is helpful to monitor the train with an accurate result. The block diagram makes the continuous update of the outcome even if the DGPS signal fails for some time. The shadowing signal will result in low tracking accuracy. But the accuracy will increase when the DGPS coverage is clear. The model is working as: if DGPS satellite visible data has 50 Hz output rate and low satellite visible has 80 Hz output, then finally mixed system will manage the data up to 80 Hz.

Using the mixed set of un-conditional model probabilities, the state transition trajectory parameters and its corresponding noise measurement estimation for SVE (X1) and SLVE (X2) filter models, respectively. Let us consider the innovations (Zi), innovations covariance matrix A, likelihood (A_{jj} and P_{IJ} are model priori and posterior probabilities of SVE and SLVE filter is computed. They are computed based on model probabilities from previous update and state transition matrix. The mixed model probability which estimates the trajectory dynamics are updated using each Kalman filter model.

The net probability estimation from the SVE and SLVE filter models is clubbed to update mixed model probabilities. The author summarized the design procedure followed in IMM algorithm using two-filter model.

- The constant position, velocity and acceleration model is designed in tracking train journey.
- Process noise variance matrix is calculated for SVE and SLVE respectively. The process noise covariance matrix is at low level for SVE and at high level for SLVE.
- Switching probability values are transforming from SVE mode to SLVE mode.

The equations outlined in the IMM algorithm are referred for n-models. The steps that follow from filter iteration process, initialization of interaction procedures are very helpful to obtain probability model for SVE and SLVE, respectively.

5.4 STABILITY CHECK ANALYSIS FOR DI-FILTER MODEL

For calculating stable transfer function, let us consider SVE model (M_1) with A (n) is an n_1^{th} polynomial and SLVE model (M_2) with B(n) is an n_2^{th} polynomial. The measuring transfer function will be of state controllable and observable.

Let us consider SVE model withpolynomial $S_1(n)$ and SLVE model with $S_2(n)$ polynomial. For closed-loop system bounded with initial conditions, it is possible to design controller using polynomial equation approach.

$y(n)/x(n) = \text{SLVE model with polynomial } S_2(n)/\text{SVE model with polynomial } S_1(n)$

For design process simplification, let us consider the net system transfer function as

$A(n) = S_1(n)$ and $B(n) = S_2(n)$

$H_1(n) = H(n)/A(n)$ for SVE model and $H_1(n) = H(n)/A(n)$ for SLVE model and

As $y(n)/r(n) = K_{Model} = B_J(n)/A_I(n)$ $\hspace{2cm}$ (5.1)

For state controllable plant B(n)/A(n), the system becomes,

$p(n)A(n) + q(n)B(n) = f(n)B(n)H_1(n)$ $\hspace{2cm}$ (5.2)

where p(n) and q(n) are state controllable and observable polynomials.

From the diagram, it is concluded that the resultant outcome for stable system is defined as

$u(n) = -\left[p(n)/f(n)*u(n) - u(n) + q(n)/f(n)*y(n) \right] + v(n)$

$v(n) = p(n)/f(n)*u(n) + q(n)/f(n)*y(n)$ $\hspace{2cm}$ (5.3)

$y(n)/r(n) = y(n)v(n)/v(n)r(n)$

$$= K_{Model} H_1(n)/H_1(n) = K_{Model}$$

$$v(n)/r(n) = K_{Model} H_1(n) \tag{5.4}$$

From the above methodology, we draw the conclusion that the stability will cause some improvement in the SVE and SLVE system models.

5.5 IMM ALGORITHM-BASED DECISION LOGIC TREE OF DI-FILTER

The two operating modes where the control characteristics of Di-filter like transmission channel, gain and measurement noise are selected in SVE and SLVE respectively. To analyze this parameter, the well-known IMM algorithm is considered. For the analysis purpose, the research proposes "M_1" for SVE and "M_2" for SLVE are operating modes of filter. The state target matrix and filter model data are design parameters for filtering methods.

Step 1: Initial Interaction—For the Di-filter which are operating in Satellite target Visible and target non-visible in no satellite service. Let ' I ' and ' J ' polynomial of equation for initial condition of filter is ¥i, j $\sum$ N

$$\hat{M}_{0J}(n-1|n-1) = \sum_{I=1}^{T} \hat{M}_I(n-1|n-1)\mu_{I|j}(n-1|n-1)$$

$$P_{0I}(n-1|n-1) = \sum_{I-1}^{T} \begin{bmatrix} P_i(n-1|n-1)+ \\ \left[\left\{\hat{M}_I(n-1|n-1) - \hat{M}_{0J}(n-1|n-1)\right\}x \\ \left\{\hat{M}_I(n-1|n-1) - \hat{X}_{0J}(n-1|n-1)\right\}^T \right] \end{bmatrix} \mu_{i|IJ}(n-1|n-1) \tag{5.5}$$

Step 2: Testing Pre-Process of Probability

The research proposes the integration of two models, SVE and SLVE, into a single mode and testing whether the model can work in both environments.

By mixing the model probability,

$$\mu_{i|j}(n-1|n-1) = \frac{1}{c_j} p_{ij}\mu_i(n-1)$$

where 'C' be the inbuilt co-efficient $\overline{c}_j = \sum\limits_{i-1}^{r} p_{ij} \mu_i (n-1)$

Let us consider the initial duration probabilities with respect to satellite visible and poor satellite visible modes are taken as 0.8 and 0.2, respectively. Later calculate how the filter gains are transform from one region to another. By calculating, mode transition probability

$$p_{ij} = \begin{pmatrix} 0.8 & 0.2 \\ 0.62 & 0.33 \end{pmatrix}$$

where P_{12} is chosen as 0.2, assuming that satellite visible model starts with low probability. Assume expected dwell time in mode is 4 seconds.

Step 3: Two -Mode Filtering model

When the probabilities of two modes are tested, the plant model state and measurement noise occurrence of covariance estimation is calculated. Based on standard prediction and procedure, the equations are listed as below. Let M be the two modes of environment, P be the mixed probability, A-B-C-D-E are plant state model design parameters, T-Y are output equation of plant model respectively.

$$M_J(n|n-1) = A_J(n-1)\hat{M}_{0J}(n-1|n-1) + B_J(n-1)u_J(n-1)$$

$$P_J(n|n-1) = A_J(n-1)P_{0J}(n-1|n-1) + A_J(n-1)^n + B_J(n-1)C_J(n-1)B_J(n-1)^n$$

$$\hat{M}_J(n|n) = \hat{M}_J(n|n-1) + D_J(n)y_J(n)$$

$$P_J(n|n) = P_J(n|n-1) - Y_J(n)E_J(n)Y_J(n)^n$$

$$Y_J(n) = Z(n) - \hat{Z}_J(n|n-1)$$

$$T_j(n) = B_J(n)P_J(n|n-1)B_J(n)^n + R_J(n)$$

$$Y_J(n) = P_j(n|n-1)B_J(n)^n T_J(n)^{-1} \tag{5.6}$$

For model match filter, the likelihood functions is given by

$$\text{Mat}_j(n) = \frac{1}{\sqrt{|T_J(n)|(2\pi)^{n/4}}} e^{-0.8\left[v_j(n)^n T_J(n)^{-1} Y_J(n)\right]}$$

Step 4: Total Estimation Values of State and Covariance measurement

The overall weighting factor that mixes probabilities obtained from the updated state and covariance equations.

$$\hat{M}(n \mid n) = \sum_{J=1}^{T} \hat{M}_J(n|n)\mu_J$$

(5.7)

$$P(n|n) = \sum_{J=1}^{T} P_J n|n) + \left\{\hat{M}_J(n|n) - \hat{M}(n|n)\right\}\left\{\hat{M}_j n|n) - \hat{M}(n|n)\right\}^n \mu_J(n)$$

The procedure listed above makes it easy to monitor the accuracy of the performance of train when it is traveling in SVE and SLVE.

From the experiment point of view, the research considered that the train route is empty without any disturbance from other train. Making one design assumption, that train is running in straight line without any turn-out or curved driving style and does not follow the predefined time chart. Figure 5.3 depicts the decision logic diagram of Di-Filter. Initially, the simulator will read the route map with the number of stops (stations), bridges, tunnels and also calculates the distance map. In a next step, it counts the distance and speed profile to reach the next station. In a similar fashion, it counts the total journey till the train reaches a predefined station (Table 5.1).

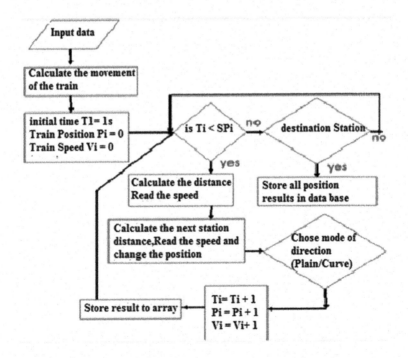

FIGURE 5.3   Decision logic diagram of di-filter.

TABLE 5.1   Presents the realistic data consider for simulation

Sampling interval (T)	1 second
Number of target states (ns)	2
Number of measurements (nm)	3
Gate probability (P_g)	0.9998
Gate size (G)	100
Distance threshold	100 m
Initial state error covariance	Identity matrix of by
Detection probability (P_d)	0.99
Process noise covariance for constant velocity model	Identity matrix of by
Process noise covariance for constant acceleration model	Not Required
Measurement noise covariance	10,000 * identity matrix of 3 by 3
Sojourn time	2
Initial mode probability	[0,9,0,1]
Constant velocity model	Satellite visible/Poor satellite mode
Constant acceleration model	Not required

By considering a unique model for the state estimation problem of train journey for different areas like SVE and SLVE, respectively. Similarly, an experiment was conducted to compare the performance of the Di-filter model used to monitor the train journey in two different environments. The models are pre-set for constant velocity and position, and the performance of each model is later compared using the IMM algorithm. Initially, the kinematic design parameters, position and velocity, are separately noted for each case when the train starts its journey from SVE to SLVE. It is assumed that the train moves along a straight track line without any disturbance. The design parameters such as state transition value, measurement noise, model gain are calculated separately for two models using the IMM algorithm. By mixing Di-filter model parameters using the IMM algorithm, it allows for good reaction, but in some cases, this combination may affect the model gain of each filter.

5.5.1 Case 1: Trajectory of Train Journey

Figure 5.4 shows the path of train movement in SVE and SLVE modes. The train starts its journey from one location in SVE to another location destiny in SLVE. The second-order kinematic measurement model is used to calculate velocity increment and process noise. For this experiment, a total of 25 scans

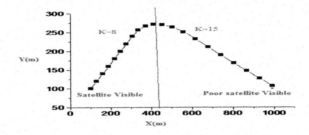

FIGURE 5.4   Trajectory of train.

(k=25) of sample data are produced with sampling duration of T=1 second. The satellite visible model for a scan of k=8 shows a velocity reading of 5 m/s for a distance of 300 and 225 m along X-direction and Y-direction respectively. Similarly, the SLVE model for a scan of k=15 shows a velocity reading of 7 m/s for distance of 600 and 215 m along the X-direction and Y-direction respectively. The trajectory spectrum of the train journey and the supporting algorithm appear to be best balanced between the two filter models.

5.5.2 Case 2: Tracking Accuracy Estimation

Figure 5.5 explains the location-based error monitoring graph of train movement in two different modes of area. The experimental results for SVE and SLVE filter models are obtained from the Kalman filter design. The experiment was carried out for 50 samples with a sampling duration of 1 second. Each filter model calculates root mean square (RMS) error for kinematic parameter. An increase in deviation and RMS error in tracking mode will make the IMM algorithm less accurate. The Kalman filter model shows less RMS error with an increase in standard deviation value (SVE) compared to the SLVE mode. Hence, the data loss in each filter model shows the way to data fusion.

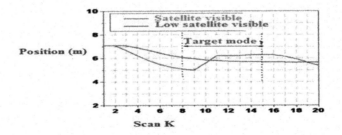

FIGURE 5.5   Position tracking accuracy estimator comparison.

5.5.3 Case 3: Tracking Train Journey Probability Model Concept

The train trajectory tracking is analyzed based on the IMM probabilistic model, which is shown in Figure 5.6. In this algorithm, model probability calculations of two filters are affected by process noise. Two filters, such as SVE and SLVE models, do selection of process noise parameters to achieve good model mix interaction. The train path shows the greatest increase in detection mode for SVE and a decrease in detection mode for SLVE probability at k=9. Similarly, the train path shows the greatest increase in detection mode for SLVE and a decrease in detection mode for SVE probability at k=15. The black curve refers to the train trajectory in SVE and gray curves refer to the train trajectory in SLVE. From Figure 5.6, it is concluded that each filter's performance depends on the state tracking matrix. When total error is present more by state noise, then largest effects are seen in SLVE.

5.5.4 Case 4: Train Trajectory Errors Estimation

The major effect of noise on the probability transition mode is clearly represented in Figures 5.6 and 5.7. When the train journey starts from SVE to SLVE, it yields better accuracy because of the measurement noise covariance

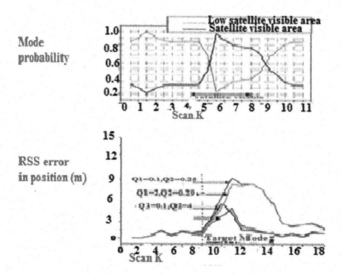

FIGURE 5.6 Probabilistic model performance in SVE and SLVE.

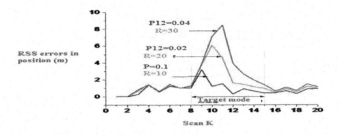

FIGURE 5.7   Transition mode with effect of onset probability.

matrix R. In SVE, the estimation error decreases much faster than in the Kalman filter. The better accuracy for R = 10 will lower the value of the tracking onset probability to $P_1 = 0.1$. Similarly, if the onset probability of $P_2 = 0.04$, then accuracy will decrease. By designing for an onset probability of $P_{12} = 0.02$, the tracking accuracy is slow while transitioning from SVE to SLVE mode. This slow period in detection on the part of the IMM algorithm will definitely increase RMS error.

The balance between the state track matrix "Q", measurement noise covariance matrix "R" and probability tracking matrix "P" are the designing factors for the selection of filters. In velocity measurement, RMS errors appeared, which in turn reflected the process noise of the filter. To build the model interaction, the detection probability is set at a minimum and the maximum will be model process noises.

Heterogeneous Sensor Data Fusion DGPS-WSN-RFID-Based Train Tracking Model

6

6.1 INTRODUCTION

Data fusion is a technique that combines samples from different number of sensors and process information. The corresponding database obtaining such information is more precise and accurate than single sensor data information. The data fusion technique is old only but advanced processing technique, usage of new sensor, and improved hardware device make fusion of data increasingly possible in real-time applications. In an intelligent vehicle navigation system, the development of sensor technology and signal processing methodology plays a major role in modern technology. By using more advanced signal-processing techniques and data fusion processes, they contribute toward the multi-sensor data fusion concept. The basic process frame set by data fusion involves fusion process at different levels namely at Interact or Identity, data level and decision level fusion.

Based on the current trend of data fusion real-time tracking applications, this chapter presents a novel data fusion process algorithm using square root information filter. The data fusion process at the decision level is done by merging information from DGPS and wireless sensors for monitoring the train journey in SVE and SLVE. The research also concentrated on multi-sensor

DOI: 10.1201/9781003294016-6

data fusion based on a probabilistic model. In this method, a single network is responsible for gathering information from many sources. The probability model is analyzed using the Bayes theorem.

6.2 RESEARCH ROAD MAP TO MULTI-SENSOR DATA FUSION TECHNOLOGY

It is an emerging technology applied to both military and civil applications. The application includes target recognition, defense surveillance system, monitoring and control of autonomous vehicle, monitoring of complicated machinery, medical diagnosis, and smart building monitoring, etc. Artificial intelligence, pattern recognition, statistical estimation is the major methodologies for the analysis of data fusion in wide range of area. Currently, data fusion systems are more familiar with tracking and detection of targets and their real-time applications. It has a past history ranging from traditional individual data collection using related techniques, to an advanced emerging engineering design.

The major areas in which the multi-sensor data fusion process can be considered for real-time application are generally DSP, statistical prediction and estimation, control theory, artificial intelligence and many more. The techniques combine data from multiple sets of traditional methods. Rapid research is needed for the development of algorithms and methods to interconnect the system. The core area of data fusion processing is Level 1 processing, where the position and velocity measurement with tracking of a single target is defined. In particular, determining the position and velocity of an object based on multiple sensor observation is a relatively old problem. The present research focuses more on calculating the correlation and maneuvering and non-maneuvering target problems. But methods such as Multiple Hypothesis Tracking (MHT), probabilistic data association method, random set theory and multiple criteria optimization theory are practiced in all sectors. Some researchers are utilizing knowledge-based systems rather than algorithm-based ones.

6.3 ARCHITECTURE AND METHODOLOGY USING DATA FUSION TECHNIQUE

The architecture selection method is a trade-off between numerical accuracy, computation time and complexity. The types of architectures suitable for data fusion are mainly classified as Centralized, Distributed and Hybrid architecture.

Centralized Architecture: In this architecture, raw data of DGPS, WSN and RFID are combined together in fusion center. The fused data is then processed and utilized in data center.

Distributed Architecture: The sensor raw data is processed before it is fed to fusion center. The fused data is directly used in data center.

Hybrid Architecture: In this architecture, processed data of DGPS - RFID and raw data of WSN are fused in fusion center. The fused data is then processed again before it is fed to the data center.

The architecture selection is completely dependent on numerical accuracy, computation time and complexity of components utilized in the respective algorithm. The decentralized architecture of Multi-Sources Data Fusion (MSDF) for proposed dissimilar sources consists of network of many sensor nodes. The sensor node has its own processing system. In this system, data fusion takes place at node basic level based on local measurements and the collected information is communicated from other nodes.

Let us consider multi-source system with two dissimilar sources S_1 (DGPS) and S_2 (WSN) with measurement data are combined at a higher level. Considering differential GPS measurements and wireless sensor measurements separately and lastly fused data from information pair and orthogonal transformation provide an enhanced numerical solution. Based on the multi-sensor observations, Information Filter state is updated continuously. We are going to define an information criterion to reject erroneous measures. The justification is intuitive: we are looking for measures that maximize information and minimize estimation error.

6.4 ARCHITECTURE FOR DGPS-WSN-BASED DATA FUSION AND MODELING ASSUMPTIONS

For detection, identification and continuously monitoring target in a predefined area is done with a ground-based surveillance system. The integrated surveillance model is in demand because of limitations in space-based radio navigation technologies. It is always welcome the integration of wireless sensor technology with a satellite navigation system to raise the tracking performance. The advantage of this technology is that it allows compensation for one another when a service failure happens. The net data obtained is improving the monitoring performance in tunnel. Thus, in a continued effort to improve

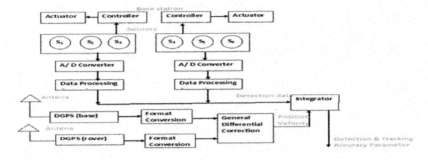

FIGURE 6.1   Block diagram of tunnel surveillance model.

tracking accuracy, we design WSN and Differential GPS-based integrating model. Figure 6.1 shows the diagram of the tracking model based on DGPS and WSN.

The data flow from DGPS measurements is analyzed using Discrete Kalman Filter Algorithm. The prediction and estimation state vector models are considered separately. The data from WSN are analyzed using Quadratic Optimal Control algorithm. The research is expected to improve the detection and tracking of train with fused information from each sensor. Information filter algorithm is proposed to analyze rail and its surrounding environments with more sensing accuracy. It is one of the advanced versions of Kalman Filter and is best suitable for multi-sensor data fusion process.

6.5 DECISION LEVEL DATA FUSION USING INFORMATION FILTER

Consider a linear system where the observations are modeled in terms of the information observed from different sources. The state vector is composed of position, velocity and acceleration in the navigation frame:

$$X = [\text{Position, Velocity, Acceleration}]$$

Let A (p, q) is an observation model where 'M' is an p-vector observation and 'X' is an q-vector of variables to be estimated. Let 'B' is a p-vector of measurement noise.

$$M = AX + B \tag{6.1}$$

By minimizing the mean square observation error, the least square solution of 'X' in terms of performance index is given as

$$J(X) = (M - AX)^T (M - AX) \tag{6.2}$$

For designing of estimation model of a linear system, the information matrix is designed in such a way that it should contain priori state information of states update. Let us consider a priori state information matrix pair,

$$IMP = (X, Y) \tag{6.3}$$

Then modified performance index based on priori state information is calculated.

$$J(X) = (M - AX)T(M - AX) + (X - X)^{TY}(X - X) \tag{6.4}$$

The algorithm to implement Information Filter is Square Root filter algorithm. This algorithm increases numerical computation accuracy and conciseness as well as fast and efficient computation. Square Root Filter (SRF) gives the square root of covariance matrix. The filter calculates the sensitivity, covariance and final solution of parameter. The algorithm avoids the inverting of small matrices sequentially formed at each time interval by substituting the inverse of one big matrix one time. The information matrix is factored and orthogonal transformation matrix is considered in equation (6.3).

$$J(X) = (M - AX)^T (M - AX) + \left(X - X^\wedge\right)^T Y^{\wedge T} Y^\wedge \left(X - X^\wedge\right) \tag{6.5}$$

$$[Y = Y^{\wedge T} Y^\wedge]$$

The equation (6.1) can be written as

$$M^\wedge = Y^\wedge X + B^\wedge \tag{6.6}$$

We observe that the performance function of equation (6.4) shows

$$J(X) = (M - AX)^T (M - AX) + \left(M^\wedge - Y^{\wedge X}\right)^T \left(M^\wedge - Y^\wedge X\right) \tag{6.7}$$

The composite system for equation (6.1) is,

$$M^\wedge = Y^\wedge X + B^\wedge$$

$$\begin{pmatrix} M^\wedge \\ M \end{pmatrix} = \begin{pmatrix} Y^\wedge \\ Y \end{pmatrix} X + \begin{pmatrix} B^\wedge \\ B \end{pmatrix}$$ (6.8)

If the priori and post-state information data is added with the fundamental observation model, then the solution of equation becomes,

$$T\begin{pmatrix} Y_{K-1}^\wedge & M_{K-1}^\wedge \\ A_K & M_K \end{pmatrix} = \begin{pmatrix} Y_K^\wedge & M_K^\wedge \\ 0 & E_K \end{pmatrix}$$

$$T\begin{pmatrix} Y_{K+1}^\wedge & M_{K+1}^\wedge \\ A_K & M_K \end{pmatrix} = \begin{pmatrix} Y_K^\wedge & M_K^\wedge \\ 0 & E_K \end{pmatrix}$$ (6.9)

The process is repeated for different measurements to get a recursive information filter.
The data-fused equation is drawn from equation (6.9),

$$T\begin{pmatrix} Y_{K+1}^\wedge & M_{K+1}^\wedge \\ A_{1K/2K} & M_{1K/2K} \end{pmatrix} = \begin{pmatrix} Y_K^\wedge & M_K^\wedge \\ 0 & E_K \end{pmatrix}$$

The fusion process obtained from two sensors such as DGPS and WSN,

$$M^\wedge FUSION(DGPS) = M_1^\wedge + M_2^\wedge$$ (6.10)

$$Y^\wedge FUSION(WS) = Y_1^\wedge + Y_2^\wedge$$ (6.11)

The data fused state is obtained by using

$$X_{FUSION}^\wedge = M_{FUSION}^\wedge(DGPS)Y_{FUSION}^{\wedge-1}(WS)$$ (6.12)

The above data equation of information filter and orthogonal transformation gives an enhanced numerical picture. We impose conclusion on fusion results as:

- Multi-source data sensor fusion will provide better information than single source.
- Fused data observation has improved reliability thus make easy for user to detect, recognize and identify target.
- The fused data should maintain all relevant data contained in the source information.
- Data losses as well as noise are suppressed in fused results to the maximum extent.

The relationship between the Performance Index and the Control input plot shows the effect of DGPS-WSN-based controller. Experimental analysis is performed using Mat lab for two values of PI and control inputs.

With control input U (K) = −0.7913 to −0.2 and PI J = 0.5–1.5, the kinematics parameter of the moving train is noted easily. Figure 6.2 describes DGPS-WSN-based Kinematic measurements [Position and Velocity] of the locomotive journey in X-direction and Y-direction respectively. The increment in PI value will lead to maximize the response. The destination of position measurement of train with WSN is not closer with DGPS measurement but the journey path considered in

By considering the same optimal control input and Performance Index value, Figure 6.3 indicates a velocity measurement graph of a moving train

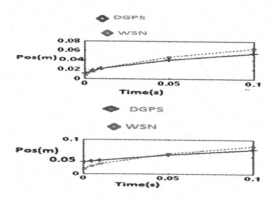

FIGURE 6.2   Position measurement along X-Y direction.

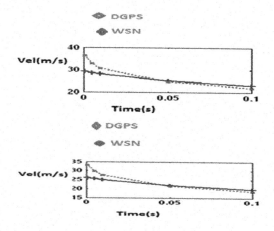

FIGURE 6.3   Velocity measurements along X-Y direction.

with DGPS and WSN measurements. From the plot, it is evident that for any state of control input,

Figure 6.4 shows the position measurement of the locomotive along X and Y directions using Square Root Information Filter Algorithm (SRIFA). The data is simulated for two dissimilar sources. Similarly, Figure 6.5 shows the

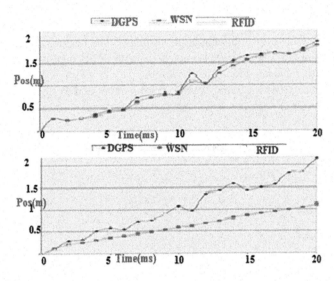

FIGURE 6.4 Data fusion position measurement of locomotive along X-Y direction.

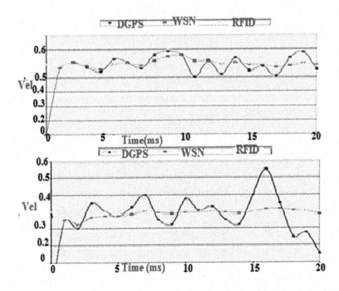

FIGURE 6.5 Data fusion velocity measurement of locomotive along X-Y direction.

TABLE 6.1   State error of DGPS, WSN and its fused data

NO OF TRAILS	KINEMATICS PARAMETERS	DGPS	WSN	FUSED ESTIMATION	FUSED MEASUREMENT
Observation I	Position (m)	0.2039	0.2113	0.2076	0.5321
	Velocity (m/s)	0.0164	0.0284	0.0224	0.0174
Observation II	Position (m)	0.2039	0.1736	0.1887	0.5310
	Velocity (m/s)	0.0164	0.0427	0.02955	0.0351

velocity measurement of the train along X and Y directions using SRIFA. It concludes that these results show the applicability of SRIFA to the problem of sensor fusion either at the data level or at the state vector estimate level and are found to be performed.

Table 6.1 shows the state error for two cases. The state errors are compared both at estimation and fusion levels.

Individual sensor will not transfer whole information than multi-sources. The fused observation is utilized to detect, recognize and identify train. The advanced version of the Kalman filter is Information Filter is proposed here and implementation is done by SRIFA. Two sets of observations for position and velocity fused parameters are verified. Finally, we conclude that the state error of position at the estimation level is less than at the measurement level. But the state error of velocity measurement at the estimation level is more than at the measurement level.

Wireless Locomotive Real-Time Surveillance Model

7

7.1 INTRODUCTION

Many sources used in the data fusion process are considered as distributed accuracy-based systems. Generally, it differs from individual source data processing in a number of methods. Typically, the unwanted sample and measurement noise are suppressed in order to obtain precise, stable with accurate output. Hence, the multi-source fusion algorithms are applicable mostly for automation applications.

In principle, the automated data fusion process is also useful where a certain amount of measurement and data are clubbed to give knowledge-based autonomous problem formulation with sufficient precision and integrity. For detection of train in SVE and SLVE continuously, the research proposed real-time Wireless Surveillance Integration model is depicted in Figure 7.1.

The DGPS-WSN-RFID-based wireless surveillance model is designed. The following are some of the interesting notes why the research is proposing this kind of technology. The DGPS gives precise tracking accuracy in locomotive identifying and detection with position, velocity measurement. The tracking accuracy will increase with an increase in the number of receivers. In coordination with the base station, the roving and stationary receivers will

DOI: 10.1201/9781003294016-7

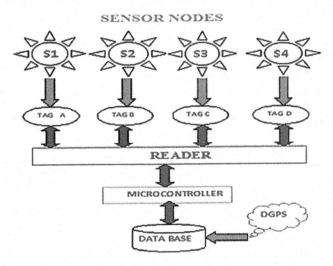

FIGURE 7.1   Block diagram of wireless surveillance model.

determine satellite measurements. In WSN, the sensor nodes are capable for detecting and monitoring the moving objects in a distribution fashion. To overcome the limitation of WSN, RFID is used for the detection and identification of locomotives. The RFID tag has memory that stores all submissive data and passes that signal to reader for processing through a wireless module which is attached to tag.

7.2 PROBABILISTIC DETECTION SENSOR DATA FUSION IDENTIFICATION MODEL

Individual sensor will not transfer whole information than multi-sources. The fused outcome of sample collection with preservation of all related data in the single-source data is really meaningful. The fusion can be of combining data from similar or dissimilar sources and later developed using various methodologies. The methods and algorithms that work on this technology are not dependent on source parameters. The major area where sensor fusion algorithms can play a major role can be classified into sections such as,

- **Least Squares Techniques of Data Fusion:** Kalman Filter algorithm

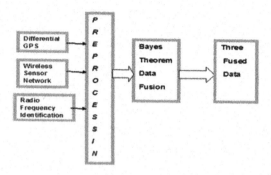

FIGURE 7.2   Flow diagram of data fusion system.

- **Probabilistic Models on Data Fusion:** Bayes theorem, Particle Filter algorithm
- **Smart Fusion:** fuzzy logic-based algorithm

The development of the sensor model is based on the state and observation process. The use of a probabilistic and information-theoretic method for fusion modeling is quite interesting. Probability density theory explains that basic probability models are responsible for fusing information with different architectures in managing sensor samples. The theorem which combines prior observation with observed data and designing matrix called probabilistic is analyzed by Bayes theorem.

Based on real-time application, channel probability may change from one sensor to another. The research proposes a DGPS-WSN-RFID-based data fusion system where fusion takes place at a higher level. Accurate estimation of train position and speed profile depending on the type of environment is required for automation. The smart rail safety system will replace the existing one.

The DGPS-WSN-RFID-based multi-sensor data fusion model is shown in Figure 7.2. In pre-processing, the input data obtained from DGPS, WSN and RFID are processed separately and then converts into a readable format by the user.

7.3 BAYES THEOREM-BASED ALGORITHM FOR DATA FUSION

It is required to calculate the train movement profile, where detection probability model is designed from three-individual sensors such as DGPS [D_1],

WSN [D_2] and RFID[D_3]. The data are obtained from train directly on sensors W_1, W_2, and W_3 respectively. The observation chart is done based on observed information distribution. Let D_1 will receive dependent observations R_1R_2, and D_2 receive observation R_2R_3, and D_3 receive observation R_3R_1 respectively. The DGPS, WSN and RFID model as three independent sources with corresponding joint probability density function P(W, D, R) is represented where W, D and R are random variables. The Probability Density Function (PDF) is represented as,

$$P(W, D, R) = P(W, D/R)P(R) \tag{7.1}$$

$$P(W, D, R) = P(W/D, R)P(D/R)P(R) \tag{7.2}$$

$$P(W, D/R) = P(W/R)P(D/R) \tag{7.3}$$

The density function is re-arranged in pattern as,

$$P(W/R) = P(R/W)P(W) \div P(R) \tag{7.4}$$

$$P(D/R) = P(R/D)P(D) \div P(R) \tag{7.5}$$

where P (W) and P (D) are Prior -Probability Detection Density Function (PDDF), the Posterior distribution function as P (W/R), P(D/R) and P (R/W), P(R/D) are Conditional PDDF respectively.

The Bayes theorem explains the relationship between prior and posterior density variable for observed information. Let P(R/W) and P (R/D) are considering as probabilistic models. The following assumptions are made while building the probability detection density sensor model.

• By maintaining the value of W = w with variable r, then PDDF is the outcome result.
• By fixing the value of D = d with r, the probability detection density function is the outcome result.
• Consider sensor model observation in the form of R = r. Finally check PDDF variation on W and D.

The Bays theorem refers to data fusion process because P(R/W) and P(R/D) are designed on the basis of fixed variable W and D with the observed information in R is defined.

Consider DGPS, WSN and RFID sensor set of observation,

$$\mathbf{R^n} = \{r1 \in R1......r_n \in Rn\} \tag{7.6}$$

Based on the construction of a pre-posterior distribution P(W| Rn) and P(D/R$_n$), the relative likelihood of values w∈W and d∈D are defined.

$$P\left(w|R^n\right) = P\left(R^n|w\right)P(w) \div P\left(R^n\right)$$
$$= P\left(r_1\ldots\ldots \mathbf{rn}|w\right)P(w) \div P\left(r_1,\ldots\ldots, r_n\right)$$
(7.7)

$$P\left(d|R^n\right) = P\left(R^n|d\right)P(d) \div P\left(R^n\right)$$
$$= P\left(r_1\ldots\ldots \mathbf{R_n}|d\right)P(d) \div P\left(r_1,\ldots\ldots.r_n\right)$$
(7.8)

The mixed distribution [P(w/r) and P(d/r)] with mixed observation P (r/w) and P(r/d) are the real values w ∈ W and d ∈ D. The data sample from ith polynomial with primary source is not depend on of another source.

On random variables w, d, r, applying the chain rule to joint PDDF,

$$P\left(w/d, r\right) = P\left(w/r\right), P\left(d/w, r\right) = P\left(d/r\right) \quad \text{it conclude that}$$

$$P\left(r|w, r1,\ldots, ri-1, ri+1, \ldots, rn\right) = P\left(r_i|w\right)$$
(7.9)

$$P\left(r/d, r_{1\ldots\ldots ri-1}, r_{i+1}\ldots\ldots\ldots r_n\right) = P\left(r_I/d\right)$$
(7.10)

With conditional independence of PDDF,

$$P\left(w, d/r\right) = P\left(w/r\right)P\left(d/r\right)$$

$$P\left(r1,\ldots,rn|w\right) = P\left(r1|w\right)\ldots P\left(rn|w\right)$$
$$= \Sigma_{i=1} P\left(r_I/w\right)$$
(7.11)

$$P\left(r1,\ldots,rn|d\right) = P\left(r1|d\right)\ldots P\left(rn|d\right)$$
$$= \Sigma_{i=1} P\left(r_I/d\right)$$
(7.12)

Replace (7.11) in equation (7.7)

$$P\left(w/r^n\right) = \left[P\left(r^n\right)\right]^{-1} P(w)\Sigma_{i=1} P\left(r_I/w\right)$$
(7.13)

Replace (7.12) in equation (7.8)

$$P\left(d/r^n\right) = \left[P\left(r^n\right)\right]^{-1} P(d)\Sigma_{i=1} P\left(r_I/d\right)$$
(7.14)

The final state likelihood values with posterior distribution w and disproportional to the product of prior and single-source likelihood. The conditional distribution P(Rn) is constant. The solution in (7.13) and (7.14) shows the direct computation methods for the relative likelihood state from many observations. For a function of w, d, and r, the P $(r_i|$ w) and P (ri/d) are priori functions with observed information sequence $Rn = \{r1, r2, \ldots, r_n\}$.

The observation sequence $\{r1, r2 \ldots r_n\}$, posterior distribution P (w/r_n) and P (d/r_n) with function of w and d is product of likelihood values with prior information P(w) and P(d) respectively.

For n-sources, the distribution information is defined,

$$P(r1,\ldots, r_n) \neq P(r1)\ldots P(r_n), \tag{7.15}$$

Each source of information is depending on the common state w ∈ W and d ∈ D. By combining all pre-post information, P (rn | w) and P (rn/d) is represented as,

$$P(w, \mathbf{Rn}) = P(w \mid \mathbf{Rn})P(\mathbf{Rn})$$

$$= P(rn, Rn - 1|w)P(w)$$

$$= P(rn|w)P(Rn - 1|w)P(w) \tag{7.16}$$

$$P(d, Rn) = P(rn/y)P(Rn - 1/d)P(d) \tag{7.17}$$

Balancing the equation on LHS and RHS,

$$P(w|\mathbf{Rn})P(\mathbf{Rn}) = P(rn|w)P(Rn - 1|w)P(w) \tag{7.18}$$

$$= P(rn|w)P(w|Rn - 1)P(Rn - 1) \tag{7.19}$$

$$P(\mathbf{Rn})/P(Rn - 1) = P(rn|Rn - 1)$$

$$P(w|\mathbf{Rn}) = P(rn|w)P(w|Rn - 1)P(rn|Rn - 1) \tag{7.20}$$

Particularly, $P(d/Rn) = P(rn/d)P(w/Rn - 1)P(rn/Rn - 1) \tag{7.21}$

The importance of solution shown in equation (7.21) will prove advancement in designing complexity and storage because it contains full information.

7.4 PERFORMANCE ANALYSIS OF DGPS-WSN-RFID-BASED MODEL

The research proposes three sources such as DGPS, WS and RFID models for two kinds of environments such as SVE and SLVE. The model analysis is described by the likelihood matrix where observation and prior information are estimated in discrete form. For simplicity, the research proposes single state **t** modeling which takes three values simultaneously.

- **t1:** t is type 1-Tracking train in SVE.
- **t2:** t is type 2-Tracking train in SLVE
- **t3:** No train journey

The corresponding possible observation information depending on t is defined as,

- **r1:** Observation of a type 1-Train is in satellite visible area
- **r2:** Observation of a type 2-Train is in low satellite visible area.
- **r3:** No train is observed.

Construct a probabilistic model based on the above observation of sensor measurement.

	P(Z1/DGPS)	P(Z2/WSN)	P(Z3/RFID)
DGPs	0.45	0.1	0.45
WSN	0.1	0.45	0.45
RFID	0.45	0.45	0.1

Posterior likelihood states of values in DGPS are measured as,

$$P(DGPS|r1,r1) = QP_{12}(r1,r1|DGPS)$$

$$= QP1(r1|DGPS)P2(r1|DGPS)$$

$$= Q \times (0.45, 0.45, 0.1) * (0.45, 0.1, 0.45)$$

$$= (0.6924, 0.1538, 0.1538)$$

From the DGPS measurements observations r1 and r2, the posterior values are estimated as (0.488, 0.488, 0.024). Using WSN shows data loss in detection performance for the same number of observations. The combined likelihood matrix for different observation in calculation for each z1, z2 observation pair is shown below.

7.4.1 Case 1: R1 = r1 and r2 = r1 r2 r3

	P(Z1/DGPS)	P(Z2/WSN)	P(Z3/RFID)
DGPs	0.6924	0.1538	0.4880
WSN	0.1538	0.6924	0.4880
RFID	0.1538	0.1538	0.0240

7.4.2 Case 2: r1 = r2 and r2 = r1 r2 r3

	P(Z1/DGPS)	P(Z2/WSN)	P(Z3/RFID)
DGPs	0.6924	0.1538	0.4880
WSN	0.1538	0.6924	0.4880
RFID	0.1538	0.1538	0.0240

7.4.2 Case 3: r1 = r3 and r2 = r1 r2 r3

	P(Z1/DGPS)	P(Z2/WSN)	P(Z3/RFID)
DGPs	0.1084	0.0241	0.2647
WSN	0.0241	0.1084	0.2647
RFID	0.8675	0.8675	0.4706

It concluded that the integration process will improve the probability detection in both type 1 target (Train in SVE) and type 2 targets (Train in SLVE). Hence, the design is excellent in the detection process and mixed sample data from sources provides improvement in overall system performance.

The experimental result shows the comparison between measurement of moving train authorities such as position and speed profile by an individual sensor and DGPS-WSN-RFID-based fused result. For simplicity the simulation pattern is arranged as follows:

1. All single sensors such as DGPS, WSN and RFID models are scaled for five evenly spaced values.
2. The data fusion model (DGPS-WSN-RFID) also scaled of five evenly spaced values.
3. The observation model is varied from +20 to −20 set of values.
4. Sensing ranges are considered from 1 to 50 and are tested feasibly.
5. Communication radius duration is varied from 10 to 45 ms in increment of 5.

For each model, ten experiments were run on the test route. The experiment shows the results of no train journey with no input or no graphs connected. When calculating the total measurement, then unconnected graph readings are not considered. The train is in SVE journey mode, Figure 7.3 depicts the prior probability detection density function. For the first set of observation up to 5 seconds, the model output is close to the initial value.

While the train takes its journey in tunnel, Figure 7.4 represents the posterior probability detection density function. The area of sensor opted for two cases is equal.

The graph of kinematic parameters (velocity and position) measurement values of train detection with different models is shown in Figure 7.4. With the variation in communication range, the location area of sensor is stable for duration up to 50 ms. The detection of train is monitored through the individual DGPS, WSN and RFID models, respectively. The bottom pattern in the graph shows RFID, as shown in Figure 7.5. The figure also describes how the sensor measurements collide with each other in the lower band of frequency using DGPS and WSN models. This is because of the detection range and high probability density. Hence, it performs good results in high detection accuracy.

As the time interval increases, the tracking graph of the RFID model reaches optimal. For t=0 to 50, it is to be noticed that the tracking accuracy graph of the DGPS model is deviated in positive PDDF, and WSN model is deviated toward negative PDDF. But the RFID model shown is neither more positive nor more negative PDDF. For t=0–20 ms, DGPS, WSN and RFID

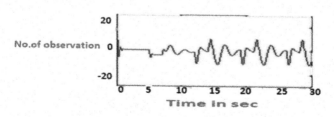

FIGURE 7.3   PDDF observation of train journey in SVE.

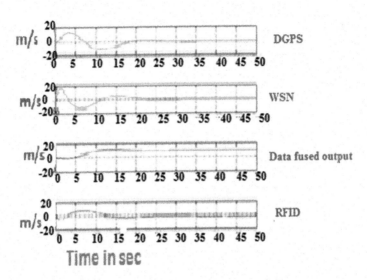

FIGURE 7.4   Position measurement of moving train.

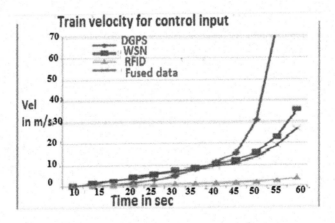

FIGURE 7.5   Velocity profile of train journey.

model the detection performance accuracy graph is almost equal. For t=20–50 ms, the accuracy of detection performance of three models is stable. But the integrated graph of three models show excellent performance for t=20–50 ms.

When train journey starts from SVE to SLVE, the velocity measurement performance by various sensor models such as DGPS model, WS model, RFID

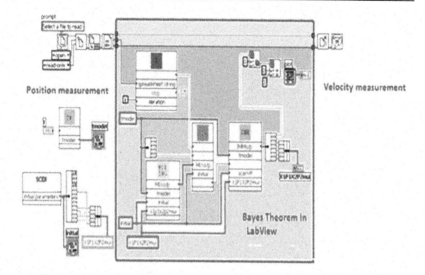

Position measurement

Velocity measurement

Bayes Theorem In LabView

FIGURE 7.6   Lab View-based data fusion.

model and three sensor-fused model are depicted in Figure 7.6 For t=0–35 ms, the DGPS and WSN model depicts the result based on PDF but RFID model shows less performed value. At t=45 ms, the integrated graph shows more accurate and stable outcome. Finally, the computed data shows 80%–90% detection accuracy based on priori and posterior error variance.

The analysis noticed that the train tracking accuracy performance is typically precise for fusion model than the measurement of individual sensor model. Hence it is concluded that there is better detection of train in SVE and SLVE by fusion model only.

Predictive Analysis of Intelligent Rail Trip Detection Service Using Machine Learning

8

8.1 INTRODUCTION

The term Artificial intelligence (AI) was coined in 1956 by John Mc-Charty in Dartmouth conference. He defined AI as 'the science and engineering of intelligent machines' in the sense that AI is a technique of getting machines to work and behave as humans. AI is accomplished by creating machines and robots that are used in various fields including healthcare, marketing, robotics, etc. AI is the science of getting machines to mimic the behavior of humans. Machine Learning is a subset of AI that focuses on getting machine to make decisions by feeding the data. Deep Learning is a subset of Machine learning that uses the concept of neural network to solve complex problems. Therefore AI, Machine learning and Deep learning are interconnected. Machine learning and Deep learning aid AI by providing algorithms and networks to solve

DOI: 10.1201/9781003294016-8

data-driven problems. Traditional Machine Learning is useful for making decisions for structured data. When the data is not structured in rows and columns or when there is a large amount of data, then finding the relationships or patterns manually is difficult. Deep learning is automatically finding out the features which are important for classification using an artificial neural network stacked layer-wise. Machine learning requires less computing power; deep learning typically needs less ongoing human intervention. Deep learning can analyze images, videos, and unstructured data in ways machine learning can't easily do. Machine learning algorithms are often easy to decode how they worked but on the other hand, deep learning algorithms are nothing but a black box. The output of traditional machine learning is usually a numerical value like score or classification. Whereas, the output of a deep learning method can be score, an element, text, speech, etc.

The data increases the performance of deep learning algorithms compared to traditional machine learning algorithms in which the performance almost saturates after a while even if the data is increased. Deep learning uses a multi-layered structure of algorithms called neural networks. The design of neural network is based on the structure of the human brain. Whenever we receive new information, the brain tries to compare it with known objects. The same concept is also used by deep neural networks. Neural networks can be taught to identify patterns and classify different types of data like the human brain. The individual layers of neural networks can also be thought of as a sort of filter that increases the likelihood of detecting and outputting a correct result.

8.2 NEURAL NETWORKS ARCHITECTURE

A neural network generally consists of a collection of connected units or nodes or nodes neurons. These artificial neurons loosely model the biological neurons of the human brain. A neuron is simply a graphical representation of a numeric value. Any connection between two artificial neurons can be considered as an axon in a real biological brain. The connections between the neurons are realized by so-called weights, which are also nothing more than numerical values. When an artificial neural network learns, the weights between neurons are changing and so does the strength of connection weights. Given training data and particular task such as classification of numbers, we are looking for certain set weights that allow the neural network to perform the classification. The set of weights is different for every task and every dataset. The prediction of values of these weights in advance is done, but the neural network has to learn them. The process of learning is called as training.

The major components in describing neural network architecture include

Input Layer: As the name suggests, it accepts inputs in several different formats provided by the programmer.

Hidden Layer: The hidden layer presents in-between input and output layers. It performs all the calculations to find hidden features and patterns.

Output Layer: The input goes through a series of transformations using the hidden layer, which finally results in output that is conveyed using this layer. The artificial neural network takes input and computes the weighted sum of the inputs and includes a bias. This computation is represented in the form of a transfer function.

$$\sum_{i=1}^{n} Wi*Xi+b$$

It determines that the weighted total is passed as an input to an activation function to produce the output. Activation functions choose whether a node should fire or not. Only those who are fired make it to the output layer. There are distinctive activation functions available that can be applied to the sort of task we are performing.

Connection Weights: Weights are the learnable parameters of a machine learning model. Weight is the parameter within a neural network that transforms input data within the network's hidden layers. When the inputs are transmitted between neurons, weights are applied to the inputs along with bias the resulting output is passed to the next layer in the neural network. The weights of a neural network are contained within the hidden layers of network. Weights control the signal (or the strength of the connection) between two neurons.

Learning rate: It is a hyperparameter that controls how much to change the model in response to estimated error each time the model weights are updated. It is challenging as a value too small may result in a long training process. The amount that the weights are updated during training is referred to as step size or "learning rate." The learning rate is a configurable hyperparameter used in the training of neural networks that has a small positive value, often in the range between 0.0 and 1.0. Learning rate determines how fast weights (in the case of a neural network) change.

Epoch: Epochs is the number of times a learning algorithm sees the complete dataset. An epoch means training the neural network with all the training data for one cycle. A forward pass and backward pass together are counted as one pass. Epochs should not overfit or underfit the data for that purpose only the required number of epochs should be given for the machine to learn. The more the number of epochs, the more the parameters are adjusted thus resulting in a better performing model. Too many epochs might lead to overfitting. If a model is overfitted, it does well in the train data and performs poorly on the test data.

Activation Function: It is used to determine the output of neural network like yes or no. It maps the resulting values between 0–1 and –1 to 1 etc. (depending upon the function).

The activation functions can be divided into two types:

1. Linear activation function
2. Nonlinear activation functions

8.3 MACHINE LEARNING METHODS

Neural learning is a process by which neural network adapts itself to stimulus by making proper parameter adjustments, resulting in the production of the desired response. Two kinds of learning are **Parameter Learning:** connection weights are updated and **Structure Learning:** change in network structure. The process of modifying the weights in the connections between network layers with the objective of achieving the expected output is called training a network. The machine learning components are classified as,

Representation: How to represent knowledge. Examples include decision trees, sets of rules, instances, graphical models, neural networks, support vector machines, model ensembles and others.

Evaluation: The way to evaluate candidate programs (hypotheses). Examples include accuracy, prediction and recall, squared error, likelihood, posterior probability, cost, margin, entropy k-L divergence and others.

Optimization: The way candidate programs are generated is known as the search process. For example combinatorial optimization, convex optimization, constrained optimization.

This is achieved through supervised learning, unsupervised learning and reinforcement learning.

8.3.1 Supervised Learning

It is the machine learning task of learning a function that maps an input to an output based on example input-output pairs. It infers function from labeled training data consisting of a set of training examples. In supervised learning, each example is a pair consisting of an input object (typically a vector) and desired output value. A supervised learning algorithm analyzes the training data and produces an inferred function, which can be used for mapping new examples. An optimal scenario will allow for the algorithm to correctly determine the class labels for unseen instances. This requires the learning algorithm to generalize from the training data to unseen situations.

8.3.2 Classification

In machine learning, classification is a supervised learning concept which categorizes set of data into classes. The most common classification problems are – speech recognition, face detection, handwriting recognition, document classification, etc. The goal is to predict the categorical class labels (discrete, unordered values, group membership) of new instances based on past observations. The majority of practical machine learning uses supervised learning. Supervised learning is where you have input variables (x) and an output variable (Y) and you use an algorithm to learn the mapping function from the input to the output $Y = f(X)$. The goal is to approximate the mapping function so well that when you have new input data (x) then you can predict the output variables (Y) for that data.

8.3.3 Line Regression

A linear regression line has an equation of the form $Y = a + bX$, where X is the explanatory variable and Y is the dependent variable. The slope of the line is b, and a is the intercept (the value of y when $x = 0$).

8.3.4 Unsupervised Learning

The type of machine learning that looks for previously undetected patterns in a data set with no pre-existing labels and with minimum of human supervision. In contrast to supervised learning that usually makes use of human-labeled data, unsupervised learning, also known as self-organization allows for modeling of probability densities over inputs. It forms one of the three main categories of machine learning, along with supervised and reinforcement learning. Semi-supervised learning, a related variant, makes use of supervised and unsupervised techniques. It is an algorithm used to draw inferences from datasets consisting of input data without labeled responses. The most common unsupervised learning method is cluster analysis, which is used for exploratory data analysis to find hidden patterns or grouping in data.

8.3.5 Clustering

The Machine Learning technique involves the grouping of data points. In theory, data points that are in the same group should have similar properties or features, while data points in different groups should have highly dissimilar properties or features. The goal of clustering is to determine the intrinsic (natural) grouping in a set of unlabeled data. It is unsupervised learning (i.e. there is no output variable to guide the learning process) data is explored by algorithms to find pattern. Connectivity models: for example, hierarchical clustering builds models based on distance connectivity. Centroid models: for example, the k-means algorithm represents each cluster by single mean vector.

8.4 PROCESSING OF DATA AND ANALYSIS

Data analysis is the science of analyzing raw data in order to make conclusions about that information. Many of the techniques and processes of data analytics have been automated into mechanical processes and algorithms that work over raw data for human consumption. This information can then be used to optimize processes to increase the overall efficiency of a business or system. It is a process of inspecting, cleansing, transforming, and modeling data with the goal of discovering useful information, suggesting conclusions, and supporting decision-making. Data analytics allows us to make informed decisions and to stop guessing. Data analysis is a broad term that encompasses many diverse types of data analysis.

Any type of information can be subjected to data analytics techniques to get insight that can be used to improve things. Data analysis is a process of inspecting, cleansing, transforming and modeling data with the analysis has multiple facets and approaches, encompassing diverse techniques under a variety of names, and is used in different business, science, and social science domains. In today's business world, data analysis plays a role in making decisions more scientific and helping businesses operate more effectively. The process of data analysis includes Data requirements, Data collection, Data processing, Data cleaning, Exploratory data analysis, Modeling and algorithms, and Data Product. Figure 8.1 shows the block diagram of data processing and data analytics.

Reality: The data is collected by day-to-day practices.

Collection of Data: Data collection is the process of gathering and measuring information on variables of interest, in an established systematic fashion that enables one to answer stated research questions, test hypotheses, and evaluate outcomes. The input data collected includes the details of the 'Date/Time', 'Latitude', Longitude', and 'Base'.

Data Selection: A data sample from different rail services is selected, often by a random selection method. In this selection process, each rail of the group stands an equal chance of being chosen as a participant in the study.

Data Filtering: It is the process of choosing a smaller part of data set and using that subset for viewing or analysis. The dataset of rail trip is so huge that a small subset of data is used to view the graphical form of result.

Data Processing: A series of actions or steps performed on data to verify, organize, transform, integrate, and extract data in an

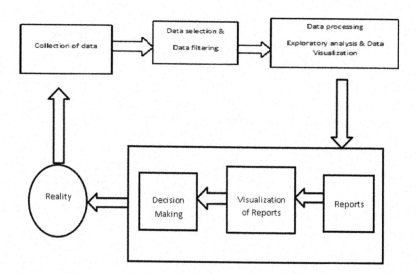

FIGURE 8.1   Block diagram of data processing and data analytics.

appropriate output form for subsequent use. It will gain insights into the areas that need improvement, decrease the chances of errors; improve accuracy and data management become less time-consuming as well.

Exploratory Data Analysis (EDA): It is an approach to analyzing data sets to summarize their main characteristics, often with visual methods. A statistical model can be used or not, but primarily EDA is for seeing what the data can tell us beyond the formal modeling or hypothesis testing task.

Data Visualization: It is the process of putting data into chart, graph, or other visual format that helps inform analysis and interpretation.

Report: A report is a document that presents information in an organized format for a specific purpose.

Visualization of Report: The visualization of the report has been done graphically so that the analysis becomes more accurate, easier, crip and clear.

Decision-Making: Process of making choices by identifying decision, gathering information and assessing alternative resolutions. The decision of day of a month or a week at which there is highest frequency of rail trip data can be known along with the prediction.

Reality: The decision made is implemented in our day to day for better performance.

8.5 ANALYSIS OF DATA SET USING RECURRENT NEURAL NETWORK

The neural network can be used for both supervised and unsupervised learning. They are very useful when it comes to certain sequential machine learning tasks, such as speech recognition. The type of neural network where the output from previous steps is fed as input to the current step. In traditional neural networks, all the inputs and outputs are independent of each other, but in cases like when it is required to predict the next word of a sentence, the previous words are required, and hence there is a need to remember the previous words. Thus, RNN came into existence, which solved this issue with the help of a hidden layer. The main and most important feature of RNN is Hidden state, which remembers some information about a sequence. Suppose there is a deeper network with one input layer, three hidden layers and one output layer. Then like other neural networks, each hidden layer will have its own set of weights and biases, let's say, for hidden layer 1 the weights and biases are (w1, b1), (w2, b2) for second hidden layer and (w3, b3) for third hidden layer. This means that each of these layers is independent of each other, i.e. they do not memorize the previous outputs. Hence, these three layers can be joined together such that the weights and bias of all hidden layers are the same in a single recurrent layer.

The obtained data set will be,

	Date/Time	Lat	Lon	Base
0	4/1/2014 0:11:00	40.7690	-73.9549	B02512
1	4/1/2014 0:17:00	40.7267	-74.0345	B02512
2	4/1/2014 0:21:00	40.7316	-73.9873	B02512
3	4/1/2014 0:28:00	40.7588	-73.9776	B02512
4	4/1/2014 0:33:00	40.7594	-73.9722	B02512
...	...	...	...	...
564511	4/30/2014 23:22:00	40.7640	-73.9744	B02764
564512	4/30/2014 23:26:00	40.7629	-73.9672	B02764
564513	4/30/2014 23:31:00	40.7443	-73.9889	B02764
564514	4/30/2014 23:32:00	40.6756	-73.9405	B02764
564515	4/30/2014 23:48:00	40.6880	-73.9608	B02764

564516 rows × 4 columns

The above data set is very huge, it consists of 564,516 rows. This may be quite difficult to do the trip analysis. In order to make the analysis quite clear and accurate a part of data set is used by passing a command in the language we code. So, the obtained data set will be

	Date/Time	Lat	Lon	Base
564511	4/30/2014 23:22:00	40.7640	-73.9744	B02764
564512	4/30/2014 23:26:00	40.7629	-73.9672	B02764
564513	4/30/2014 23:31:00	40.7443	-73.9889	B02764
564514	4/30/2014 23:32:00	40.6756	-73.9405	B02764
564515	4/30/2014 23:48:00	40.6880	-73.9608	B02764

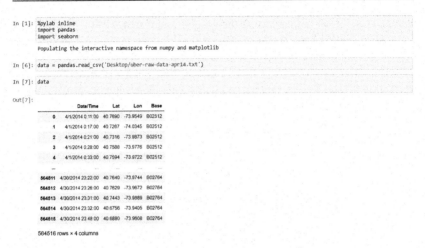

In the above code, pandas are imported to do the data manipulation and sea-born to view the graphical plot. The rail trip data which acts as an input has been imported to do the data analysis.

The data is in tabular form which includes Date/time, Latitude, Longitude, Base. Date/Time: Drop and pick up time. Lat and Lon: Are the positions.

Base is the rail number.

The data set obtained is very vast, in order to view the results accurately a sub-set of data is considered, hence data cleaning (data selection and data filtering) process is used.

In order to obtain the first five rows of data set a code data. head () is executed, and for the last five rows a code data. tail () will be executed. In order to add a days of month column to the dataset, pandas are mapped to ('Date/Time').

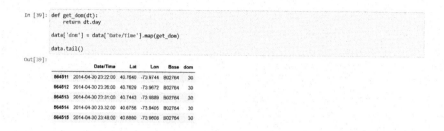

The addition of some useful columns such as days of month (DoM) is done in the above code because to analyze the data and to draw the outcome of it will be easier. Since there are 30 days in month column DoM results as 30.

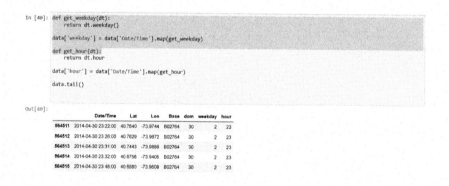

The same analysis will be done for weekday and for hours in a day.

The above graph Figure 8.2 is drawn frequency v/s date of the month. The x-axis is labeled as' date of month' and y-axis as 'Frequency'. The frequency is highest at the 30th day and value is found to be 36,251. This indicates that the no. of trips done by the particular rail is high in the 30th day as compared to the other days.

In order to know the exact frequency values, from the obtained graphical plot the above code is used.

```
In [41]: hist(data.dom, bins=30, rwidth=.8, range=(0.5, 30.5))
         xlabel('date of the month')
         ylabel('frequency')
         title('Frequency by DoM - uber - Apr 2014')

Out[41]: Text(0.5, 1.0, 'Frequency by DoM - uber - Apr 2014')
```

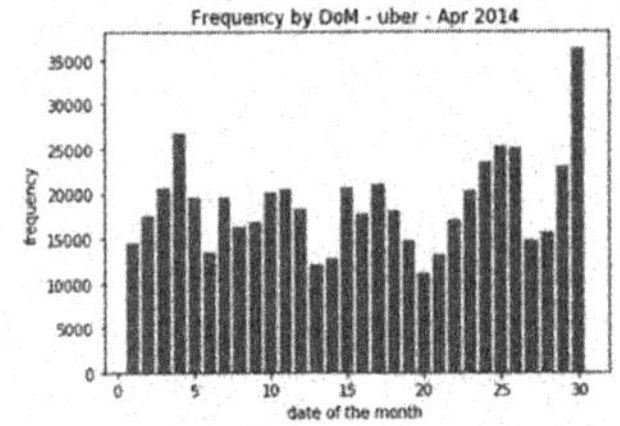

FIGURE 8.2   Graph of frequency vs. date of month.

```
In [42]: #for k, rows in data.groupby('dom'):
         #    print((k, len(rows)))

         def count_rows(rows):
             return len(rows)

         by_date = data.groupby('dom').apply(count_rows)
         by_date
```

```
Out[42]: dom
         1      14546
         2      17474
         3      20701
         4      26714
         5      19521
         6      13445
         7      19550
         8      16188
         9      16843
         10     20041
         11     20420
         12     18170
         13     12112
         14     12674
         15     20641
         16     17717
         17     20973
         18     18074
         19     14602
         20     11017
         21     13162
         22     16975
         23     20346
         24     23352
         25     25095
         26     24925
         27     14677
         28     15475
         29     22835
         30     36251
         dtype: int64
```

```
In [44]: by_date_sorted = by_date.sort_values()
         by_date_sorted

Out[44]: dom
         20    11017
         13    12112
         14    12674
         21    13162
          6    13445
          1    14546
         19    14602
         27    14677
         28    15475
          8    16188
          9    16843
         22    16975
          2    17474
         16    17717
         18    18074
         12    18170
          5    19521
          7    19550
         10    20041
         23    20346
         11    20420
         15    20641
          3    20701
         17    20973
         29    22835
         24    23352
         26    24925
         25    25095
          4    26714
         30    36251
         dtype: int64
```

The sorting of DoM has been done in ascending order based on the frequency values of Days of month as depicted in Figure 8.3 The 20th day of the month has the lowest frequency with a value 11,017 and 30th day of the month has the highest frequency value of 36,215. The sorted frequency values have been analyzed with a graphical plot as shown in Figure 8.3.

The analysis of hour has been done as same as DoM. There are 24 hours in a day hence in x-axis 'the no of hours in a day' is labeled in the scale of 24 and y=axis is labeled as 'frequency'. An hour with the highest frequency denotes that the rail has been in trip service for longer duration.

The analysis of days in a week has been done as same as DoM. There are 7 days in a week, hence in x-axis denotes the name of days in a week i.e., 'Mon – Sat' and y-axis is labeled as 'frequency' as mentioned in Figure 8.4.

```
In [45]: bar(range(1, 31), by_date_sorted)
         xticks(range(1,31), by_date_sorted.index)
         xlabel('date of the month')
         ylabel('frequency')
         title('Frequency by DoM - uber - Apr 2014')
         ;

Out[45]: ''
```

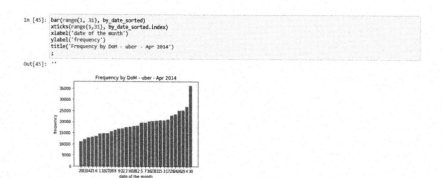

FIGURE 8.3   Graph of frequency v/s no. of hours in a day.

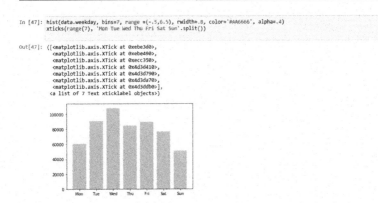

```
In [47]: hist(data.weekday, bins=7, range =(-.5,6.5), rwidth=.8, color='#AA6666', alpha=.4)
         xticks(range(7), 'Mon Tue Wed Thu Fri Sat Sun'.split())

Out[47]: ([<matplotlib.axis.XTick at 0xebe3d0>,
          <matplotlib.axis.XTick at 0xebe490>,
          <matplotlib.axis.XTick at 0xecc350>,
          <matplotlib.axis.XTick at 0x4d3d410>,
          <matplotlib.axis.XTick at 0x4d3d790>,
          <matplotlib.axis.XTick at 0x4d3da70>,
          <matplotlib.axis.XTick at 0x4d3ddb0>],
         <a list of 7 Text xticklabel objects>)
```

FIGURE 8.4   Graph of frequency v/s days of week.

A day of week with the highest frequency denotes that the rail has been in trip service for longer duration. The cross-analysis of week day and the hour has been done using a heat map in order to know at which particular day and corresponding time the frequency was high.

The latitude at which the frequency is high is spotted so that the prediction of the demand for that particular location can be done. It also explains how the relationship between latitude and frequency is high and how it can be spotted so that the prediction for the demand for that particular location can be done. As shown in the above graph Figure 8.4, the data collected is analyzed in various ways, i.e. by month, week, hour, longitude and latitude. The graphs show peaks and dips of the trips taken by rail on different days, at different times. Keeping all this in mind, we can make any changes if and when required. This process is made easier due to the machine learning algorithm used, as has been stated in the beginning.

Further Readings

L. Aguado. "A Low Cost Low Power GPS Positioning System for Monitoring Landslide." In: *NAVI Tech*, 2007.

S. Antonov, A. Fehn, and A. Kugi. "Unscented Kalman Filter for Vehicle State Estimation." *Vehicle System Dynamics*, vol. 49, no. 9, pp. 1497–1520, September 2011.

Association of American Rail-Roads. "Positive Train Control System" pp. 1–4, April, 2013. https://www.aar.org/keyissues/Documents/Background.

Patrik Axelsson and Fredrik Gustafsson. "Discrete-Time Solutions to the Continuous-Time Differential Lyapunov Equation with Applications to Kalman Filtering." *IEEE Transaction on Automatic Control*, vol. 60, no. 3, pp. 632–643, 2014.

P. Bennett. "Wireless Sensor Networks for Underground Railway Applications: Case Studies in Prague and London." *Smart Structures and Systems*, vol. 6, no. 5/6, pp. 619–639, 2010.

A. Budiyono. "Principles of Optimal Control with Applications." Lecture Notes on *Optimal Control Engineering*, Department of Aeronautics & Astronautics, Bandung Institute of Technology, 2004.

Y.T. Cheg and B.S. Chen, "Fuzzy Approach for Robust Reference Tracking Control Design of Nonlinear Distributed Parameter Time Delayed Systems and its Applications." *IEEE Transactions on Fuzzy Systems*, vol. 18, no. 6, pp. 1041–1057, 2010.

W. Chen and Y. Fu. "Cooperative Distributed Target Tracking Algorithm in Mobile Wireless Sensor Network." *International Journal of Control Theory and Applications*, vol. 9, no. 2, pp. 155–164, 2011.

Y. Chen and C. Han. "Maneuvering Vehicle Tracking Based on Multi-Sensor Fusion." *Acta Automatica Sinica*, vol. 31, pp. 625–630, 2005.

T. Cho, C. Lee, and S. Choi. "Multi-Sensor Fusion with Interacting Multiple Model Filter for Improved Aircraft Position Accuracy." *International Journal on Sensors*, vol. 13, no. 4, pp. 4122–4137, 2013.

Grace S. Deaecto, Matheus Souza, and Jose C. Geromel. "Networked Control Systems." *IEEE Transaction on Automatic Control*, vol. 56, no. 5, pp. 423–431, 2013.

H. Dong, B. Ning, and B. Cai. "Automatic Train Control System Development and Simulation for High-Speed Railways." *IEEE Circuits and Systems Magazine*, vol. 10, no. 2, pp. 6–18, 2010.

Farhad Farokhi and Karl H. Johansson. "Model Information for Discrete-Time Linear Systems with Stochastically-Varying Parameters." *IEEE Transaction on Automation Optimal Control Design*, vol. 60, no. 3, pp. 877–881, 2014.

M. A. Hannanet. "Intelligent Bus Monitoring and Management System." *IEEE Vehicular Communication Journal*, vol. 21, pp. 235–241, 2012.

Mark Hartong, Rajni Goel, and Duminda Wijesekera. "Positive Train Control (PTC) Failure Modes." *Journal of King Saud University – Science*, vol. 23, pp. 311–321, 2011.

Will Hedgcock. High Accuracy Difference Tracking of Low Cost GPS Receiver. Elsevier, 2011.

Y. Hou. "Multiple-Sensor Fusion Tracking Based on Square-Root Cubature Kalman Filtering." *Journal of Network*, vol. 9, no. 7, pp. 1955–1961, 2014.

Indian Railway Annual Report. "Safety Works on Indian Railways – Anti Collision Device (ACD) and Train Protection and Warning System (TPWS)." Report No. 32 of 2011-12.

A. Kurian. "Data Fusion in an Intelligent Transportation Systems: Progress and Challenges—A Survey." *Journal of Information Fusion*, vol. 12, pp. 4–10, 2011.

X. Lieu and A. Goldsmith. "Wireless Communication Tradeoff in Distributed Control." In: *Proceedings of the 42nd IEEE Conference on Decision and Control*, pp. 682–694, December 2009.

C. Luo. "Multi Sensor Fusion and Integration: Approaches, Applications and Future Research Directions." *IEEE Transaction on Data Fusion*, vol. 23, no. 4, pp. 566–578, 2001.

R. Merwe, A. Doucety, N. Freitas, and E. Wan. "The Unscented Kalman Filter Advances." In: *Neural Information Processing Systems*, Vancouver, Canada, 2000.

Bin Ning. "A Comprehensive Information System for Railway Network." *Computers in Railways VIII*, WIT Press, pp. 152–162, 2003, Comp Rail 2002, Lemnos, Greece.

Bin Ning. "ETCS and CTCS." *Computers in Railways IX*, WIT Press, pp. 262–272, 2004, Comp Rail 2004, Drensdon, Germany.

Bin Ning. "Intelligent Railway System in the 21st Century." *Computers in Railways VII*, WIT Press, pp. 1153–1163, September, 2000, Comp Rail 2000, Bologna, Italy.

B. Ning, T. Tang, H. Dong, D. Wen, D. Liu, S. Gao, and J. Wang. "An Introduction to Parallel Control and Management for High-Speed Railway Systems." *IEEE Transaction on Intelligent Transportation System*, vol. 12, no. 4, pp. 1473–1483, December 2011.

B. Ning, T. Tang, and K. Qiu. "CTCS-Chinese Train Control System." *WIT Transactions on the Built Environment*, vol. 74, no. 15 pp. 1–7, 2008.

Tanuja Patgar and Ripal Patel. "Comparative Analysis of face Recognition using Deep Learning." In: National Conference on recent Trends in Electrical, Instrumentation, Electronics & Communication Engineering, 2019.

Tanuja Patgar, Ripal Patel, and S. Girija. "Real Conversation with Human-Machine 24/7 COVID-19 Chatbot Based on Knowledge Graph Contextual Search." *Data Science and Computational Intelligence*, vol. 1483, pp. 258–272, 2021.

Tanuja Patgar and Shankaraiah. "A Heterogeneous Access Remote Integrating Surveillance Heuristic Model for a Moving Train in Tunnel." *International Journal of Intelligent System and Application*, vol. 8, no. 3, pp. 59–65, 2016.

Tanuja Patgar and Shankaraiah. "A Kinematics Update State Hypotheses Information Surveillance Model for a Moving Train." *International Journal of Electronics and Electrical*, vol. 7, pp. 30–36, 2015.

Tanuja Patgar and Shankaraiah. "Performance Analysis of Multi-Source Data Fusion Tracking Algorithm for Ground Based Surveillance Model to Monitor the Moving Locomotive." *International Journal of Advances in Computer and Electronics Engineering*, vol. 1, no. 3, pp. 23–30, 2016.

Tanuja Patgar and Shankaraiah. "The Impact of Hybrid Data Fusion Based on Probabilistic Detection Identification Model for Intelligent Rail Communication Highway." *International Journal of Sensor and its Applications for Control System*, vol. 4, no. 2, pp. 9–20, 2016.

Tanuja Patgar and Shankaraiah. "Trajectory Tracking of Locomotive Using IMM-Based Robust Hybrid Control Algorithm." *International Journal of Sensors and Sensor Networks*, vol. 5, no. 3, pp. 34–42, 2017.

Tanuja Patgar and Triveni. "CNN Based Emotion Classification Cognitive Model for Facial Expression." *Turkish Journal of Computer and Mathematics Education*, vol. 12, no. 3, pp. 6718–6739, 2021.

H. Qi and J. B. Moore. "Direct Kalman Filtering Approach for GPS/INS Integration." *IEEE Transactions on Aerospace and Electronic Systems*, vol. 38, no. 2, pp. 687–693, 2002.

Duncan Smith and Sameer Singh. "Approaches to Multi-Sensor Data Fusion in Target Tracking: A Survey." *IEEE Transaction on Knowledge and Data Engineering*, vol. 18, pp. 1696–1710, 2016.

Y. Sun, S. Zang, H. Xu, and S. Lin. "Cooperative Communication for Wireless Ad hoc Sensor Network." *International Journal of Distributed Sensor Network*, vol. 2013, Article ID 161268, 2 pages, 2013.

Z. F. Syed, Priyanka Aggarwal, X. Niu, and N. Ei-sheimy. "Civilian Vehicle Navigation: Required Alignment of the Inertial Sensors for Acceptable Navigation Accuracies." *IEEE Transactions on Vehicular Technology*, vol. 57, no. 6, pp. 30402–30412, 2008.

Sebastian Thrun, Yufeng Liu, Daphne Koller, Andrew Y. Ng, Zoubin Ghahramani, and Hugh Durrant-Whyte. "Simultaneous Localization and Mapping with Sparse Extended Information Filters." *International Journal of Robotics Research*, vol. 23, pp. 693–716, 2004.

U.S. Coast Guard Navigation Centre Report. "NAVSTAR GPS User Equipment Introduction." August 1, 2011.

David Waboso. "Progress on the Industry Plan for ERTMS Implementation in the UK." *The Proceedings of Aspect 2003*, London, UK, pp. 251–252, September 2003.

J. H. Wang and Y. Gao. "Land Vehicle Dynamics-Aided Inertial Navigation." *IEEE Transactions on Aerospace and Electronic Systems*, vol. 46, no. 4, pp. 1638–1653, 2010.

S. S. Wang. "Multi-Feature Fusion Tracking Based on a New Particle Filter." *Journal of Computers*, vol. 7, pp. 2939–2947, 2012.

X. Yang, M. Xu, and Y. Cheng. "Risk Analysis for the Modifications in Automatic Train Control System." In: *Proceedings of the 2010 International Symposium on Computer Communication Control and Automation (3CA)*, Tainan, China, May 5–7, 2010.

H. Zhang and Y. Zhao. "The Performance Comparison and Analysis of First Order, Second Order EKF, Smoother for GPS/DR Navigation." *Optik*, vol. 122, pp. 777–781, 2011.

L. Zhang and Z. Wang. "Integration of RFID into Wireless Sensor Networks: Architectures, Opportunities and Challenging Problems." In: *Proceedings of the Fifth International Conference on Grid and Cooperative Computing Workshops* (GCCW'06), 2006.

H. Zhao, T. Xu, and T. Tang. "Towards Modeling and Evaluation of Availability of Communications Based Train Control (CBTC) System." In: *Proceedings of the IEEE International Conference on Communications Technology and Applications Communications Technology (ICCTA)*, Beijing, China, October 16–18, 2009.

W. Zou and W. Sun. "A Multi-Dimensional Data Association Algorithm for Multi-Sensor Fusion." *Intelligent Science and Intelligent Data Engineering*, vol. 25, pp. 280–288, 2013.

A. Zulu and S. John. "A Review of Control Algorithms for Autonomous Quad Rotor." *Journal of Applied Sciences*, vol. 4, pp. 547–556, 2014.

Index

Note: **Bold** page numbers refer to tables; *italic* page numbers refer to figures.

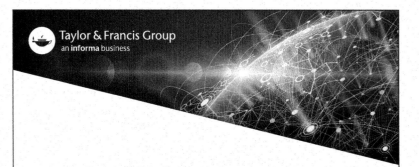